SMOKEY'S
BEST DAMN GARAGE
IN TOWN

NORTON
16
GOODYEAR
CAM2
SUNOCO
CHAMPION
MONROE
ARCA

CHEVY STOCK CARS

Dr. John Craft

MBI Publishing Company

First published in 2000 by MBI Publishing Company, 729 Prospect Avenue, PO Box 1, Osceola, WI 54020-0001 USA

MBI Publishing Company books are also available at discounts in bulk quantity for industrial or sales-promotional use. For details write to Special Sales Manager at Motorbooks International Wholesalers & Distributors, 729 Prospect Avenue, Osceola, WI 54020-0001 USA.

Cover Design: Dan Perry
Layout Designer: Todd Sauers

Library of Congress Cataloging-in-Publication Data
Craft, John Albert.
Chevy stock cars / John Craft.
p. cm. — (Enthusiast color series)
Includes index.
ISBN 0-7603-0587-0 (pbk. : alk. paper)
1. Chevrolet automobile. 2. Stock car racing—United States. 3. NASCAR (Association) I. Title. II. Series.
TL215.C6C76 2000
629.228—dc21 99-44191

On the front cover: The most-feared Chevrolet of all time just may be Dale Earnhardt's black #3 Monte Carlo. *John Craft*

On the frontispiece: Chevrolet Impalas sprouted wings for 1959. *John Craft*

On the title page: The big fender of the mid-1970s car body style provided for plenty of close quarters action on the NASCAR tour's short-track bullrings. The car's long beak and short bustle also offered optimal weight distribution. *John Craft*

On the back cover: Rex White and the #4 Chevrolet played a prominent role in NASCAR in the late 1950s and early 1960s, with White winning the title in 1960. *John Craft*

Printed in China

CONTENTS

Coca-Cola
BUCK BAKER
OWNER BOBBY GRIFFIN
DARLINGTON INTERNATIONAL RACEWAY
GRIFFIN MOTORS
87
AIR LIFT
PURE
CHAMPION
MONROE
Firestone

Chapter

THE 1950s

First Blood for the General

General Motors (GM) and Chevrolet have held a dominant position in stock car racing circles for so long that many think it has always been this way. And that's very easy to understand when one considers that since the very first season of Winston Cup (formerly Grand National) competition, the "General's" cars have visited victory lane more than 750 times. Competition cars of the Bow Tie persuasion have fared best of all over the first 50 years of National Association for Stock Car Auto Racing (NASCAR) fender-rubbin', accounting for more than 450 of those wins. Pontiac drivers have racked up more than 120 stock car trophies, and their Oldsmobile counterparts 116 more. Buick stockers round out the tally with 65 circle track triumphs. But GM drivers didn't always have a lock on stock car glory; in fact, it took Chevy drivers the best part of a decade to score their first Grand National (GN) win.

Look closely, this probably was your father's Oldsmobile. Especially if he happened to be a NASCAR race car driver in 1950. Rocket 88s like the one Buck Baker drove in the early 1950s were the cars to beat at the time. *Mike Slade*

1949 and 1950: Rocket 88s Roar to Life

The racing world was a far different place in 1949 (when Big Bill France first cooked up the NASCAR series) than it is today. As originally conceived, for example, the series was based on American-built sedans that were required to be unchanged in any way from showroom trim. France, in fact, called NASCAR's top division the "Strictly stock series." And that name, and the restrictions it carried with it, was the reason for Chevrolet's initial lack of racing success. Though a 1948 Chevy had paced the Indy 500 the year before the NASCAR series was formed, in strictly stock trim a sleepy little 90-horsepower stovebolt six cylinder was anything but a race-ready powerplant. That probably explains the total absence of Chevrolets on the starting grid at the very first NASCAR race in Charlotte in June 1949.

But that didn't mean that General Motors drivers had to sit on the sidelines during that first season of competition. Not by a long shot. In fact, then, as now, some of the fastest cars on the starting grid at the old Charlotte fairgrounds had genuine GM pedigrees. In 1949 those cars were literally your father's

Big Bill France originally called his premier racing division the strictly stock series. And one look into the cockpit of Buck Baker's 1950 Rocket 88 Olds is all it takes to understand why. Things were disconcertingly stock.

The secret to the Rocket 88's success was a free-revving, 303-cubic-inch, overhead-valve, small-block engine. Though modest in performance by modern standards, a Rocket 88 engine struck fear into the hearts of all who had to race against it.

Oldsmobile. Olds ad flacks called the cars in question Rocket 88s, and, for once, Madison Avenue hype actually came pretty close to the truth. The cars truly were rocket ships.

The secret of the cars' speed was the all-new overhead valve-equipped (big news in the late Cretaceous), 303-cubic-inch V-8 engines that Olds introduced in 1949. Though rated at a mere 202 horsepower, the all-new engine had most of its late 1940s rivals covered in the "motorvation" department, even when installed in a not-so-svelte 3,455-pound Olds 88 with a full-frame chassis.

Race number one in NASCAR history featured no fewer than six Rocket 88 drivers. In that number were future series stars Tim Flock and Red Byron. Byron was an Atlanta-based mechanic who had built his reputation by souping up both 'shine cars and the revenuer pursuit vehicles that nightly played cat and mouse on the "thunder roads" of the rural Southland. Flock sprang from the driving side of the 'shine business having learned the finer points of nocturnal high-speed driving (that is, fleeing and attempting to elude) from his two older 'shine-hauling brothers, Bob and Fonty.

Though neither Byron nor Flock finished first that day in Charlotte (Byron came closest with a third-place berth), it didn't take long for a GM-badged comp car to visit victory lane. Not long at all, in fact. Oldsmobile/General Motors win number one came in the second race of the inaugural 1949 season in July. That particular race took place in a sleepy little Florida beach town called Daytona that would soon figure large in stock car racers' aspirations everywhere.

As longtime NASCAR fans recall, that first "Cup" race took place on a 4.15-mile-long beach and road course that incorporated equal stretches of paved portions of Highway A1A and white sandy (low tide!) beach straightaways connected by two fairly tight berm-flanked turns. Qualification in those days did not (as today) consist of hot laps around the track, but rather involved measured speed runs down the beach and back. When the sand had settled, Gober Sosbee's Rocket 88 qualified fastest overall with Red Byron's Olds just a few ticks of the stopwatch slower.

The race itself took 2 hours, 33 minutes to complete and turned out to be a Rocket 88 romp. Sosbee and Byron battled for the lead much of the event until Byron put him away for good on lap 34. Oldsmobile drivers Tim Flock, Frank "Rebel" Mundy, and Joe Littlejohn followed Byron's #22 across the line to make the finish a sweep for the Olds division. It wouldn't be the last time that a #22 General Motor's stocker crossed the finish line first at the race in Daytona.

All told, Oldsmobiles scored five wins in the eight-race 1949 Strictly Stock season. Byron's two victories and four top-five finishes earned him the modest sum of $5,800 and the season championship. Had there been a manufacturer's championship in 1949, Byron would have claimed that for the Olds division also.

The secret to the Rocket 88's success was a free-revving, 303-cubic-inch, overhead-valve, small-block engine. Though modest in performance by modern standards, a Rocket 88 engine struck fear into the hearts of all who had to race against it.

Bill Blair won the 1953 Daytona Beach race in a block-long Oldsmobile just like this one. He had the good fortune that day to be in the right place at the right time when race leader Fonty Flock's Rocket 88 ran out of gas on the last lap. Blair also had the good fortune to be driving a 135-horsepower Olds 88. A replica of Blair's car is currently on display at the North Carolina Auto Racing Hall of Fame in Mooresville.

Bill Rexford and two soon-to-be-famous fellows named Curtis Morton Turner and Glenn "Fireball" Roberts picked up the Oldsmobile banner and ran with it the very next season. Though Turner ultimately scored more 88 wins than Rexford, the NASCAR points system that, even then, rewarded consistency more than wins made Rexford the second series champ. Roberts also finished ahead of Turner in the points chase, though he only had one "X" in the win column. Both Roberts and Turner would go on to visit victory lane many more times in the upcoming Grand National seasons.

1951–1954: Other Manufacturers Meet the Challenge

The wins scored by Olds drivers in 1949 and 1950 on the NASCAR circuit did not go unnoticed. Sportier lights in the buying public quickly recognized the performance potential of the Rocket 88 car line and began to show up in Olds showrooms in increasing numbers. That fact made an impact in Detroit. With an eye toward bolstering their own sales, sales types at Hudson decided to steal some of the glory being grabbed up by Rocket 88 drivers on the NASCAR tour. When it was determined that the flat head sixes under the hoods of race Hornets weren't up to the task of besting Olds' small-block overhead valve (OHV) V-8s, Hudson engineers were ordered to whip up some go-fast add-ons designed specifically for stock car competition. Hornet teams were able to skirt the NASCAR rules book by calling the performance hardware "export" equipment, causing tech inspectors to look the other way at teardown time. The end result was a much faster Hudson Hornet stock car that was able to battle with the best of the Olds 88 drivers.

Brothers Tim, Bob, and Fonty Flock joined forces to carry on Olds' winning ways, and as a result Rocket 88s visited victory lane circa 1951 many more times (20, in fact). But Herb Thomas' 7 wins and 15 top-five Hudson finishes (and the "export" bits and pieces under his hood) carried the day and the season championship. Fonty Flock finished up the year with 8 Oldsmobile-backed wins and was second in the points chase.

Hudson engineers redoubled their "export" engine component production for 1952. That act, coupled with the generous salaries offered to drivers by Hudson's now factory-backed racing effort, lured a number of top drivers (including Tim Flock) into the Hornet camp. As a result, the next four seasons were a struggle for General Motors drivers generally and Olds drivers in particular. One bright spot came in September of 1952 when Fonty Flock notched the first GM Southern 500 win in his Air Lift Special Rocket 88. Bill Blair bested all comers at the 1953 beach race in an Oldsmobile, and Buck Baker notched a second Southern 500 win for Olds at Darlington the same year. But those wins were the exception to the rule for the once all-conquering Rocket 88s. And things only got worse when Chrysler got into the NASCAR game.

1955: Overhead-Valve V-8s Save the Day

While Hudson and Chrysler teams were tearing up the Grand National tour during the mid-1950s, interesting things were going on behind the scenes at the various General Motors divisions. Buick engineers were hard at work making the straight eight a thing of the past. As it happens, Chevrolet engine and foundry folks were pursuing that same goal. And so, too, were their counterparts in the

The control cabin in Blair's Olds featured a bench seat (racing buckets were years away in 1953) and a not-so-sporty column shifter. The bus-sized wheel no doubt made navigating the sandy beach course at Daytona less of a chore.

Don't look for headers, four-barrel carburetors, or other modern go-fast goodies under the hood of Blair's 88. Nonetheless, the lumbering behemoth had what it took to win on Daytona's fabled beach course.

Mid-1950s Oldsmobiles were both fast and good looking, even when festooned with dirt-deflecting front screens like the one pictured here on driver Ed Negre's 1956 Rocket 88. *JDC Collection*

Pontiac division. As was the custom in the General's ranks in those days, engineers in the various divisions all worked from their own clean sheets. The days of the corporate small block were still—blissfully—decades in the future. As a result, by the end of the decade the tranquility of the Southland was regularly disrupted (on race days, that is) by the basso-profundo bark of four very different General Motors OHV V-8s. And before the nifty fifties and poodle skirts had both worn out their welcome, the world-changing Chevrolet small-block engine had exploded onto the motorsports scene.

In 1955, Chevrolet's "mouse" motor made its debut on both the showroom floor and starting grids all across the country. Work on the all-new engine commenced with the installation of Edward Nicholas Cole as manufacturing manager of the Chevrolet division. Cole's arrival on the scene coincided with the initial planning for the 1955 model year. A major part of the plan for the cars slated to take a bow that year was the development of an all-new OHV V-8 engine. Cole knew from the outset that if Chevrolet was going to have a shot at the new youth performance market, the division was going to have to develop an engine that would transform Chevy's reputation for building trustworthy but tepid transportation.

Cole tapped Ed Kelley and Harry Barr to design the new engine, and their basic design proved to be so successful that it is still in use today—both on and off the track. Working with a set of design parameters that included five main bearing journals and a desired displacement of 265 cubic inches, Kelley and Barr put pen to drafting paper and set to work. In final form the new engine was 30 percent more powerful than the sleepy little six cylinder it replaced while at the same time weighing a full 40 pounds less than the old stovebolt. PR flacks called the new 8.0:1 compression ratio mill the "Turbo-Fire" 265, and it boasted 162 horsepower when topped with a two-barrel carburetor. Kelley and Barr upped that figure to 180 before the end of the 1955 production year with the addition of a "Power Pack" option package consisting of dual exhausts and a single Carter WCFB four-barrel fuel mixer.

The new body style that the free-revving little small block was destined to serve duty in was also radically redesigned for 1955. There's little doubt today that Chevy's styling efforts that year rate right up there with the best designs of any car manufacturer anywhere at any time. But looks don't win races. Redesigned and greatly stiffened chassis, lighter ball-joint-equipped suspension components, and an 18 percent overall reduction in weight do. And

NASCAR tech inspectors circa 1955 hard at work. Unfortunately, their post-race labors after the 1955 beach race in Daytona took away Fireball Roberts' Buick win. Note the oh-so-racy white walls on Fireball's Fish Carburetor-backed car. *JDC Collection*

Buck Baker became Chevrolet's first National Driving Champion in 1957. He continued to campaign Bow Tie-based race cars like this #87 ragtop into the 1960s. *JDC Collection*

Chevy's new 150 and 210 sedans sported all those desirable traits. In due course, that translated into NASCAR victories.

Stock car win number one for Chevrolet and its all-new small-block engine came in March 1955 on a half-mile dirt track in Columbia, South Carolina. Fonty Flock was driving a #14 150 sedan for car owner Frank Christian that day and began the race from the back of the pack. Brother Tim started at the head of the field in his Kiekhaeffer-prepared Chrysler 300 and seemed destined to use his car's 300 horsepower to dominate the event. But, when the dust settled, it was older brother Fonty who got to kiss the pretty girl. That kiss and the $1,000 in prize money Fonty earned were the first sweet tastes of racing victory for Chevrolet drivers. Cars of the Bow Tie persuasion have been a regular fixture in winners' circles all across the country ever since.

Legendary mechanic and car builder Henry "Smokey" Yunick proved that Fonty Flock's first Chevrolet victory was no fluke seven months later in the Southern 500 at Darlington. The fifth running of the Southern 500 had all the ingredients of a Hollywood thriller (and one decidedly better written than Tom Cruise's less-than-realistic *Days of Thunder*). The cast of players included crusty and conniving master mechanic Smokey Yunick. Driver Herb Thomas, who had nearly been killed in a racing accident four months earlier, played the sympathetic lead. Drama was provided by Thomas' hospital bed prediction that he would bounce back from his crash in Charlotte to win a third Southern 500. Outboard motor magnate Karl Kiekhaeffer and his all-conquering fleet of Chrysler 300s (cars that had been rolling over the competition like Hitler took France) played the heavy. The character actor role was filled by Indy champ Mauri Rose, who at Smokey's behest located a number of sets of special sports car tires that Firestone had created for road race work at LeMans. The underdog role was played by Yunick's brand-new little small-block-powered "Motoramic" 150 sedan. Down on power to Kiekhaeffer's great white fleet of 300-plus horsepower

The interior of Phil Reed's 1959 Impala.

ton
305 H.P.

JIM REED
JIM REED'S GMC TRUCK SALES
PEEKSKILL, N.Y.
7
CHAMPION

Fonty Flock honed his driving skills in the north Georgia hills hauling liquid freight by moonlight. He scored Chevrolet's first win when he put his abilities to use in the Grand National ranks at Columbia, South Carolina. *Daytona Racing Archives*

Chryslers and unproven as the new kid on the block, Smokey's 1955 Chevy had to rely on brains rather than brawn to reach victory lane.

When race day dawned, Darlington was packed to the rafters with the Southern 500's first sell-out crowd of more than 50,000 spectators. As expected, when the green flag fell, Thomas' #92 Chevy was unable to claw its way to the front of the 69-car field on horsepower alone. In fact, Tim and Fonty Flock's Chrysler 300s dominated the first hundred-odd laps of the event. Until, that is, their heavier and faster cars started encountering tire trouble. By Smokey's count there were 680 blown tires during the running of the 1955 Southern 500, and most of those were suffered by cars other than Herb Thomas' little Chevy. By race's end, the special road race tires that Rose had salvaged from an Akron, Ohio, junkyard had proven to be worth their weight in gold. With Kiekhaeffer's cars far in arrears, the race was decided between Thomas' Chevy and "Little" Joe Weatherly's #9 Ford. When Little Joe's left front rim failed on lap 317, the resulting contact with the outside wall ended both a 129-lap stint in the lead and any chance of beating Thomas for the win. This would not be the last time that a Ford and a Chevrolet battled it out for stock car glory on the NASCAR tour.

In 1955 GM's Buick division also scored its first Grand National wins. The very first Buick triumph came (and went) at the Daytona Beach race in February. Rising star and hometown boy Glenn "Fireball" Roberts used his 255-horsepower Roadmaster to qualify fourth for the race and then went on to lead every lap of the 160-mile race to score what appeared to be Buick's first GN win. Unfortunately, a post-race teardown revealed pushrods that had been turned down 16/100ths of an inch. That discovery led to Fireball's disqualification 24 hours after he'd been declared the winner.

It was left to future champion Buck Baker to score Buick's first official win three months later at a 100-mile dirt track race at Charlotte Speedway. Herb Thomas claimed win number two for Buick in a Smokey Yunick car at Raleigh one month before his Southern 500 win at Darlington. As things turned out, those

Lee Petty started 17 convertible races in his "zipper" top Olds during the 1957 season. The cars (like other convertible comp cars of the day) were referred to as zipper tops because they'd race as ragtops one day and as hardtops the next. Rocket 88 Oldsmobiles were fast in both categories. *JDC Collection*

Buick drivers were around for the very first race of the all-new NASCAR series in 1949. By the mid-1950s, the Buicks were some of the fastest cars on the tour. Here, Buick pilot Pete Yow is seen dodging ruts in the 1956 beach race at Daytona. *Daytona Racing Archives*

Though Chevrolet introduced the 348-cubic-inch, big-block "W" engine in 1958, most Bow Tie racers chose to stick with their tried-and-true small-block 283s that year. It was a different story by 1960 when big blocks like the one in Junior Johnson's Kennel Club Special powered ahead and replaced the small block in Chevy racers' hearts.

two wins would be the last scored by a Buick racer in the Grand National division for 26 years. (But when Buick would eventually return to the winner's circle in 1981, it would do so with a vengeance.)

1956: Chrysler Rampage

Nineteen fifty-six dashed the hopes of GM racers everywhere (and just about everybody else too). Kiekhaeffer's Chryslers went on a tear that season and won nearly every race they entered. Surprisingly, at season's end, Kiekhaeffer suddenly closed up shop and left racing for good after winning two back-to-back Grand National championships. Chevrolet got serious about stock car racing in the fall of 1956 by setting up the Southeastern Engineering Development Company (SEDCO) in Atlanta. SEDCO was devoted exclusively to running Bow Tie stock car operations. Once it was up and running, Chevy racers' fortunes began to improve.

When the 1957 season got under way, for the first time in two seasons all bets were off, and just about anybody had a chance to win.

1957: AMA Ban Silences Strong Ford Competition

Chevrolet took steps to make sure that more than a few of those "anybodys" were Bow Tie-mounted drivers by introducing a new and improved (read—bigger and more powerful) version of the small-block engine. Following the time-honored maxim that bigger is better, Chevy engineers broke out the boring bar and expanded their OHV engine's internals to 283 cubic inches. Better yet, they also cast up an all-new fuel-injected induction system that pushed horsepower figures to the magical 1-pony-per-cubic-inch level. When installed in the newly redesigned 1957 Chevy line, a formidable race car was created.

The only flies in the ointment were those pesky engineers over at Ford who had some induction tricks of their own up their sleeves. Like their Chevrolet counterparts, Ford folks had an all-new overhead valve V-8 to tinker with in the mid-1950s, and the quest for more grunt led them to bolt on a belt-driven McCulloch supercharger that put 1957 Fords out ahead of their "fuelie" competition both under the hood (300 horsepower to a 283's 283) and on the track.

Ford's factory-backed drivers (including a fellow named Ralph Moody) began the 1957 season with a string of victories that were sure to have thrown a wet blanket on Chevy's hope for headlines. Ford drivers won 15 of the first 21 events contested in the Grand National division in 1957. Chevy drivers, on the other hand, took the checker in just 5 races. The one bright spot in the statistics was the first Pontiac win in the NASCAR ranks, turned in by Cotton Owens at Daytona in February. Owens' win was convincing, as he led all but one of the race's 39 laps on the combination beach/road course. The power put out by the Ray Nichels-prepped Pontiac V-8 under Owens' hood proved that engineers in the Indian-head division were on the right track with their own engine development program. But, as mentioned, any promise displayed in Owens' beach race triumph was overshadowed by Ford's overall dominance on the track. Something would have to be done—and quickly!

The man with the plan turned out to be GM exec Harlow "Red" Curtice. Ford's dominance at the track in early 1957 put both Chevrolet and General Motors in a tough spot. Though the Bow Tie division had engineers like Zora Duntov who were literally chomping at the bit to go racing in a big way, Curtice felt constrained in giving them their head. The reason was the corporation's fear that highly publicized factory-backed GM racing might draw too much attention to the fact that the General's market share in those days came perilously close to deserving Justice Department scrutiny for antitrust violations. Curtice decided if he couldn't defeat his Blue Oval rivals on the track, he would just have to outsmart them in the boardroom. Lucky for him that Ford's top executive at the time was an exceedingly gullible fellow named Robert McNamara.

Curtice sealed Ford's racing fate far from the tracks by convincing McNamara to sign on to the American Motorsports Association (AMA) ban on factory-backed motorsports competition that Curtice himself had cooked up. McNamara swallowed the bait—hook, line, and sinker— and totally shut down Ford's factory-backed racing teams at mid-season. The record book reflects that from that point on, Chevrolet's racing fortunes took a dramatic turn for the better. Though Bow Tie teams had won just 5 Grand National events during the first part of 1957, Chevy drivers finished first in 14 of the 32 NASCAR races that took place after the ban. In that number was a win by

Buck Baker (#87) scored Chevrolet's first Grand National driving championship in 1957 with the help of a fuel-injected, 283-powered "Black Widow." With Bud Moore turning the wrenches, Baker won 10 races and finished in the top five at 20 other events. Speedy Thompson can be seen bringing up the rear in another "Black Widow." *JDC Collection*

Bill France opened the gates to his palace of speed in Daytona in 1959. Its 33-degree banking initially took the breath away from most drivers on the tour when they first saw the track—but not for long. Soon hotshoes like Richard Petty (pictured here in a blue #43 Olds convertible) were tearing around the 2.5-mile track with reckless abandon. *JDC Collection*

Speedy Thompson in the 1957 running of the always-important Southern 500 at Darlington.

At season's end, Chevy driver Buck Baker had secured both his second and Chevrolet's first Grand National driving title. The fuel-injected 1957 "Black Widows" that team mechanic Bud Moore built for Baker produced 10 wins, 30 top-five finishes, and more than $30,000 in prize money.

While Chevrolet drivers like Baker, Jack Smith, Bob Wellborn, and Speedy Thompson were adding check marks to the Chevrolet win column, a fellow named Lee Petty from tiny Level Cross, North Carolina, made the switch from Chrysler to Oldsmobile. In his entourage was a gangly lad named Richard who helped turn wrenches on dad's 277-horsepower Super 88s. Father and sons' (the elder Petty's other son Maurice also helped out with pit chores) efforts produced four Oldsmobile wins in 1957. These weren't destined to be the last GM wins scored by a member of the Petty family.

1958: Petty Power

With archrival Ford Motor Company (Fomoco) snoozing on the sidelines and the General's R&D engineers still hard at work producing new and improved go-fast parts for Chevrolet, Oldsmobile, and Pontiac NASCAR teams, 1958 promised to be a good year for GM teams on the circuit. And indeed it was.

Future champion Rex White began the competition year with a win for Chevrolet in the first race of the calendar, and Paul Goldsmith piloted Smokey Yunick's Pontiac to a convincing win from the pole at the beach race in Daytona. Smokey's black-and-gold Poncho carried #3 racing livery and was powered by Pontiac's 370-cubic-inch answer to Chevrolet's newly introduced 348-cubic-inch "big-block" engine. Though both divisions were supposedly out of the racing business and officially honoring the AMA ban that Red Curtice had engineered, race-oriented engineers were burning the midnight oil in a not-so-clandestine quest for more and more speed.

Interestingly, though Bow Tie teams had access to the newly introduced Impala car line and its 348 "W" engine, most opted to re-up their enlistment with the tried and proven 1957 vehicles they'd raced the season before. It turned out to be a wise move. Fireball Roberts, for one, went on a tear in his 1957 car, winning the Southern 500 and five other races. Chevy pilots like Buck Baker, Speedy Thompson, and Jim Reed helped bring Chevrolet's season total to an impressive 23 wins. In addition, Pontiac drivers brought home three more wins for the GM banner that year.

Even so, these numbers were not good enough to keep Lee Petty from winning a third Grand National championship for Oldsmobile. Using the same formula that would later serve son Richard so well, Petty ran just about every one of the 51 events that made up the 1958 season and won 7 of them. Twenty-one other top-five finishes added up to Petty's second national title and $26,565 in winnings.

1959: The Debut Daytona 500

In 1959, Big Bill France opened his high-banked palace of speed in Daytona. Once he did, the face of Grand National stock car racing was forever changed. Gone was the treacherous old part-sand and part-pavement beach course, and in its place was a speedway so big and so tall that drivers were struck dumb in wonder (read—fear) the first time they pulled into the infield.

The very first Daytona 500 was run on February 22, 1959. Bob Wellborn was fastest during qualifying, and he set the first fast lap around the Big D at a then-blistering speed of 140.121 miles per hour. Unfortunately, the engine under the hood of Wellborn's 1959 Impala only lasted until lap 75 of the 200 that made up the race. The late stages of the race

During the 1950s many future stars got their start in NASCAR's convertible division, including a lanky lad named Richard Petty. He is seen here at Daytona in 1959 literally driving his dad's Oldsmobile (#43). He soon graduated to mounts with roof panels and Grand National stardom. *JDC Collection*

Following pages
Junior Johnson won the 1960 Daytona 500 in a Ray Fox-prepped Impala just like this one. Apparently, his race car was a completely stock street car until just one week before the February 14, 1960, event. Floridian Don Rhuff built this replica of Johnson's 1959 race winner.

JUNIOR JOHNSON
PURE
27
Built By
RAY FOX
PERFECT CIRCLE
NASCAR
INTERNATIONAL
DAYTONA BEACH KENNE
CARTER ELEC. CO.
Daytona Beach
EQUIPPED WITH
CHAMPIO
SPARK PLUGS

CLUB
IR LIFT
KEHOE
NICKEY
1959 348

Chevrolet Impalas sprouted wings for 1959. And that's perhaps part of the reason they flew around NASCAR's superspeedways that season. The 305-horsepower, 348-cubic-inch "W" engines under their hoods probably helped a little bit too. Here Bob Wellborn is seen battling it out with a couple of Fords in the 1959 Daytona 500. Some things never change. *JDC Collection*

saw Chevrolet drivers take a back seat to T-Bird-mounted Johnny Beauchamp and Oldsmobile pilot Lee Petty. Though Petty's block-long (and finned!) Olds 88 didn't look the part of a race car, from lap 148 forward it was the only car in the 59-car field that was able to give Beauchamp a run for his money—even after the checkered flag fell. And that's because the final dash across the stripe that Beauchamp and Petty made was so exceedingly close that at first NASCAR officials declared Beauchamp the winner by a hair. It wasn't until 61 hours later that a photo analysis of the finish revealed it had been Petty's land-yacht-sized 88 that had actually edged Beauchamp out for the win. And so it was that the first Daytona 500—the Super Bowl of stock car racing——was won by your father's (or at least Richard's father's) Oldsmobile.

Chevrolet drivers had another good year on the tour and snagged 14 Grand National trophies in the process. Jim Reed drove a "bat-winged" 1959 Impala to glory in the Southern 500, and Rex White drove a similarly winged #4 Chevrolet to 5 short-track wins. That having been said, it was Lee Petty who once again snagged the national driving title. And though he'd opened the year in an Olds, the elder Petty jumped ship and returned to Chrysler shortly after the 500, so his GN crown provided little reason to celebrate back in Detroit.

As the 1950s ended, Ford was still securely on sabbatical. The dawning of a new decade held more than a little promise for General Motors racers.

41
CURTIS TURNER
DELTA AUTO SALES
PURE
SAFETY
BELTS
49
BOB WELBORN
TUXEDO PLUMBING & HEATING CO.
PURE

REX WHITE
320 H.P.
TRUCK CENTER
EQUIPPED WITH
Gabriel
SHOCK ABSORBERS
4
PERFECT CIRCLE
WYNN'S
FRICTION PROOFING
RADIATOR ADDITIVE
powered by
PURE
Grey-Rock
BRAKE LININGS

Chapter

THE 1960s

BANNED IN BOSTON

The kick-off of the new decade had to have been an optimistic affair for GM racers in general and Chevrolet drivers in particular. The Ford Motor Company (Fomoco) was still securely on the sidelines, and Chrysler's racing efforts were primarily concentrated on the two teams fielded by Lee Petty. The odds were clearly in favor of continued GM dominance in stock car circles for the length of the new decade, or so it seemed.

The 1960s began exactly as anticipated, but they would not end that way. Few who gathered in the garage area in advance of the 1960 running of the Daytona 500 would have guessed that the 1960s would turn out to be a decade filled with lost opportunities for drivers in every one of the General's divisions. But at least that "winter of discontent" didn't descend on GM until 1963. And 1960 was a particularly good year for Chevrolet drivers in the GN ranks.

Though this man may be unfamiliar to many modern NASCAR fans, he was well known and feared on the circuit in the late 1950s and early 1960s. His name is Rex White, and during his time on the tour he finished in the top-10 at an incredible 70 percent of the races he started. And most of those events found him at the helm of a Chevrolet race car. In 1960 he won both the fans' hearts (Most Popular Driver) and the season points race (Grand National Champion). *JDC Collection*

1960: Chevy Success Continues

Junior Johnson started the ball rolling by winning the 1960 Daytona 500 in a Ray Fox-prepped Impala. Interestingly, ace mechanic Fox originally had planned to sit out the 500. All that changed just nine days before the race when a group of backers proposed that he build a car for the second running of the 500. Working around the clock, Fox was able to transform a showroom 1959 Impala into a race-winning Grand National stock car in just seven short days. Legendary driver Junior Johnson translated a ninth-place start into a first-place finish with the help of the 320 ponies cranked out by the 348 big-block engine under the hood of his bat-winged Chevrolet. Johnson's win was made more dramatic by what track announcers perceived as a series of near rear-end collisions in which he was involved. To them it appeared that Johnson's Chevrolet nearly ran into the back of a number of cars at different points in the race. What was really happening was Johnson's discovery of superspeedway drafting, and the near collisions were simply Johnson

Bubble-top Catalinas were some of the prettiest cars on and off the track in 1961. Here's a trio of them being driven at Daytona that year by (left to right) Junior Johnson, Cotton Owens, and David Pearson. *JDC Collection*

setting up the cars he nearly "crashed" into for slipstream passes. Nose-to-tail racing at Daytona is now a common sight, all because of Junior's 1960 discovery at the helm of a bat-wing Impala.

Joe Lee Johnson backed up Junior Johnson's superspeedway win with a victory in the inaugural World 600 at the newly opened Charlotte Motor Speedway, but Rex White was the most successful GM driver that season. His 6 Impala wins and 25 top-five finishes were what it took to secure the first Grand National driving title of the 1960s. His winnings for the 1960 title topped $57,000.

Pontiac drivers also proved to be a force on the 1960 tour and ultimately scored just 2 fewer wins than the 15 turned in by the Bow Tie brigade. Fireball Roberts drove Smokey Yunick's #22 to a popular win in the Dixie 300 at the new superspeedway in Atlanta, and Buck Baker notched his second Southern 500 win in a block-long Poncho of his own.

If Chevrolet drivers were concerned by the success of their Pontiac brethren, they had good reason, as they would find out just one season later.

1961–1962: Super-Duty Pontiac Power

Chevrolet's Grand National victories made quite an impact on both the buying public and a couple of fellows named Knudsen and Wangers. Both just happened to work in Chevrolet's rival Pontiac division, and both were inveterate racers. Best of all for Pontiac fans, both occupied positions in Pontiac's food chain that permitted just about every one of their high-performance whims to become actual corporate fact in a very short time (bureaucratically speaking, that is).

Semon "Bunkie" Knudsen had taken control of Pontiac's leadership in 1958, and Jim Wangers served Knudsen as the division's chief ad man. Both were dedicated to the idea that Sunday wins on the track translated

Chevrolet driver Johnny Beauchamp made headlines during Speedweeks 1961 at Daytona when he and MoPar-driver Lee Petty took a little excursion off the track during a qualifying race. Fortunately, neither driver was killed. *JDC Collection*

into Monday traffic on the showroom floor. To that end both (like their Bow tie counterparts) had made sure that the back door at Pontiac's R&D division remained wide open to racers, even though the division was officially adhering to Curtice's AMA ban on factory-backed racing.

Knudsen and Wangers' "racy" outlook, coupled with Pontiac's styling decision to downsize the Catalina line for 1961, helped make 1961 and 1962 standout seasons for drivers in the "Indian Head" division. Marvin Panch fired the first shot in Pontiac's battle to defeat archrival Chevrolet by winning the 1961 running of the Daytona 500 in a Smokey Yunick-prepared Pontiac. When Panch's gold-and-black #20 Ponch crossed the stripe, the only other cars on the same lap were Joe Weatherly's Bud Moore-built Pontiac and Paul Goldsmith's Ray Nichels-built Catalina.

Adding insult to injury was the fact that Pontiac drivers had also swept both of the twin qualifiers that preceded the 500. Fireball

Fireball Roberts
22
STEPHENS PONTIAC
Daytona Beach.
Florida

Roberts used his #22 Yunick-prepped Catalina to lead Jim Paschal and Jack Smith's Catalina to the checker in race one, and Fireball's black-and-gold car cinched the 500 pole with a 155.709-mile-per-hour lap on pole day. Qualifier two ended with a Pontiac train made up of Joe Weatherly, Marvin Panch, and Cotton Owens arriving at the line before everyone else. As you can see, Daytona Speedweeks in 1961 were most definitely an all-Pontiac affair.

Fireball Roberts took in the rapidly passing 1962 scenery from this perch. Note the essentially stock nature of Grand National stock car control cabins in 1962 as well as car builder Yunick's total lack of color coordination.

Previous pages
Fireball Roberts won the Daytona 500 in 1962 with the help of this 421 Super Duty Catalina. Smokey Yunick built the #22 car in his Daytona Beach "Best Damn Garage in Town" shop, and hometown hero Roberts went on to score the popular win. More recently, the black-and-gold car has been on display at the International Motorsports Hall of Fame in Talladega, Alabama.

And that's pretty much the same way that the rest of the season went too. Pontiac drivers like Junior Johnson, David Pearson, and Jack Smith captured almost the entire slate of headline-producing big track events in 1961. The only "big one" that got away was the Southern 500 that Nelson Stacey claimed in a Holman & Moody-equipped Ford Starliner. Even so, the rules book's favoritism (in points) for drivers that started most of the races on the tour resulted in yet another Grand National driving title for Chevrolet, this time scored by Ned Jarrett in his #11 Impala. Though the soft-spoken Jarrett won but a single race during the 50-race season (compared to 9 for Joe Weatherly and 7 for Junior Johnson), he started 46 races and scored top fives in 23 of them. And that was enough to take the title and $41,055.90 in winnings.

The big news for 1962 was the addition of a punched-out version of the Super Duty (SD) 389 to the Pontiac arsenal. First introduced in late 1961 as an over-the-counter, drag race-oriented powerplant, the new 421-cubic-inch SD Poncho motor made the Catalinas captained by Fireball Roberts, Joe Weatherly, Cotton Owens, and Jack Smith even more intimidating than they had been the year before.

Driving to Daytona for Speedweeks 1962 must have been like a condemned man's stroll toward the electric chair for Chevy drivers that year. After all, if their 409 engines hadn't been able to best Pontiac drivers powered by SD 348s the year before, how were they going to be able to beat those same drivers and their bigger and better 421 SD motors in 1962? The answer: they weren't.

Fireball Roberts made this clear during qualifying by capturing the pole with a speed of 156.999 miles per hour and then going on to convincingly win the first of Daytona's two traditional qualifying races.

Chevrolet engineers introduced a new motor at Daytona in 1963 that featured poly-angled valves, a beefy four-bolt bottom end, and generously proportioned, equally spaced ports. Developed under a veil of secrecy even within GM ranks, the new engine quickly became known as the "Mystery Motor."

Joe Weatherly completed the qualifying trifecta for Pontiac by winning his qualifier in Bud Moore's #8 Poncho. Roberts went on to dominate the race proper on his way to his first Daytona 500 win. It was an incredible and impressive performance for Pontiac for the second straight year, and Chevrolet drivers were distraught.

Fireball Roberts became the King of Daytona that year, winning the Firecracker 250 on July 4th in a Banjo Matthews prepped Catalina (having split with Smokey shortly after Speedweeks). Joe Weatherly took his first Grand National driving title in Bud Moore's red-and-black Catalina by winning 9 events and finishing in the top five at 30 others. All told, Pontiac drivers had won an incredible 52 of 105 races contested during the 1961 and 1962 seasons.

1963: Factory Ban Backfires/ Mystery Motor Unveiled

Unfortunately, Pontiac drivers were collectively destined to win just four more times in 1963 and then remain absent from NASCAR victory lanes for the next 17 straight seasons.

There are several reasons for the dramatic reversal of racing fortune that awaited Pontiac (and GM) drivers during the balance of the 1960s. Part of the problem was Henry Ford II's decision in early 1962 to formally invalidate the 1957 AMA ban on factory-backed racing that Red Curtice had snookered Robert McNamara into. (By then McNamara was long since gone from the "Glasshouse" in Dearborn,

Junior Johnson navigated superspeedway traffic with this not-so-ergonomic steering wheel in 1963. Note the production-based bucket seat and instrument panel. Chevrolet stock cars were mostly just that in 1963.

Though Roger Penske has long been associated with motorsports as a winning Indy car and NASCAR team owner, it might come as a surprise to some that during the 1960s he was one of the best sports car drivers in the country. When not at the helm of a usually Chevrolet-powered sporty car, he, on occasion, tried his luck in the stock car ranks. His #02 Pontiac is pictured here at Riverside in 1963. *JDC Collection*

having joined the Kennedy administration as secretary of defense.) Pontiac's total domination of the NASCAR (not to mention NHRA) series in 1961 and 1962 finally convinced the powers-that-be at Ford that GM wasn't honoring the AMA ban after all. That coupled with the clock cleaning that Ford dealers were getting in the race for sales convinced Henry Ford to ditch the ban and go racing. A revitalized Holman & Moody quickly set about taking wins away from both Pontiac and Chevrolet drivers shortly thereafter. Chrysler's decision to get more involved in factory-backed NASCAR racing didn't help either.

But the big reason for the hard times the GM drivers fell upon after the 1963 season was the very same AMA ban on factory-backed motorsports that had hamstrung Ford drivers since 1957. You see, Pontiac's total domination on the track had also captured the attention of GM's top brass. Their concern about possible monopoly litigation being initiated by trust busting U.S. attorneys was, if anything, greater in 1962 than it had been five years before. As a result, they decided that a little profile lowering was called for, and the best way to achieve that goal was to actually honor the AMA ban that Red Curtice had engineered in 1957. Go figure.

As a result, the funding pins were knocked out from under both Pontiac's and Chevrolet's clandestine racing programs just before the beginning of the 1963 NASCAR season. So serious was GM about shutting down its racing efforts that "repossession" orders went out to race teams just before the

Daytona 500 that were designed to reclaim any high-performance parts (and cars) that had already gone out to racers through GM's now-closed (welded shut, actually) back door. And it goes without saying that all factory-subsidized R&D work on new racing-destined components also came to a crashing halt.

It's easy to imagine the disarray that GM's decision left Pontiac and Chevrolet camps in on the eve of the most important race of the season. Even so, there were still some bright points for GM drivers in the 1963 season before the lights went out in their race shops altogether.

Chevrolet's racing program for the 1963 Grand National season was particularly promising in the days just before the Daytona 500. Stung by Pontiac's domination of the high banks (and the low ones too) in 1961 and 1962, Bow Tie engine and foundry types set out to build a race-specific big-block engine that was better able to handle the competition than the not-so-fine 409 had been.

Starting with a clean sheet of paper, they began by inking a rock-solid bottom-end design that featured a unique four-bolt main journal design consisting of conventional two-bolt caps that were then surrounded by an additional double-bolted reinforcing girdle. A battleship-strength-forged reciprocating assembly came off the drawing board next, followed by a set of free-flowing head castings that carried a poly-angle valve layout and equally spaced intake and exhaust ports. Bits and pieces of the package also included a trick cowl induction "Ram Air" setup and a pair of cast-iron exhaust manifolds that flowed just about as well as a matching pair of still-outlawed tubular headers could.

A veil of mystery surrounded the all-new engines. So much so, in fact, that people began to call them "Mystery Motors." A handful were assembled for pre-season testing, and both Pontiac and Chevrolet drivers were invited to Pontiac's southwestern proving grounds (in Mesa, Arizona) for a head-to-head comparison of the new motor with Pontiac's SD 421 engine of the season before.

In that number was North Carolinian and Pontiac racer Junior Johnson. When it became apparent that the new 427-cubic-inch Chevy motor was more than a match for the SD 421, Johnson and others (such as Smokey Yunick) elected to switch to Impalas for 1963.

Though the NASCAR series has always been the dominant stock car racing series, it wasn't always the only one. During the early days, a number of other sanctioning bodies gave Big Bill France's series a run for its money. The United States Auto Club (USAC) was in that number, and for a while in the early 1960s, stock car races were sanctioned by that body across the Midwest. Here is a shot of Roger Penske (#02) in his Ray Nichels-prepped Pontiac chasing Parnelli Jones (#15) in a USAC race at Indianapolis Raceway Park in 1963. *JDC Collection*

Following pages
Junior Johnson switched to Chevrolet for 1963 when a pre-season comparison proved that the all-new Mystery Motor that Chevrolet introduced that season was more than a match for the 421 Super Duty Pontiac motor he had campaigned the year before. Unfortunately, GM pulled the rug out from under all General Motors drivers by dropping its factory backing just before the Daytona 500 that year. *Daytona Racing Archives*

VROLET
PURE
Grey-Rock
RayFox
ENGINEERING
3
Poultry
INC.

Joe Weatherly won a second straight Grand National drivers' title in 1963 driving a string of different race cars. When his factory-backed Pontiac ride (pictured here) ran out of steam shortly after Atlanta in the spring of the season, he jumped from car owner to car owner piloting Pontiacs, Plymouths, Dodges, and even Mercurys on his way to the points title. *JDC Collection*

As mentioned, GM's decision to finally honor the AMA ban was a very belated one, and one that didn't take place until race teams were well into their preparations for Speedweeks 1963 at Daytona. As you might have guessed, GM's attempt to "repossess" every one of the 48 new MkII 427 motors (as they were officially called) that had been cast up to that date was met with little enthusiasm by the teams that had received them. Ultimately, Chevy execs decided (probably wisely) against driving to the hills of Carolina (and elsewhere) to actually try and physically take back the engines.

Even so, GM shut down all development of the engine and stopped all production of Mystery Motor engine parts cold. Though racers like Johnson and Yunick would be campaigning MkII 427-powered Impalas in 1963, they'd have to get by on only the spare parts they'd amassed before GM shut down the program.

Bobby Johns campaigned a series of #7 Chevrolets during the mid-1960s. His colorful blue-and-white cars were often the only Bow Tie-badged competition cars in the garage area. One wonders how much better he and the other independent GM drivers could have run with the same level of factory backing their Brand X rivals enjoyed. *JDC Collection*

Who said that the new Taurus race car is the first four-door racer to compete in the NASCAR series? Anyone who makes that claim surely has forgotten the #6 four-door 1964 Olds that Ed Brown drove in the 1965 Motor Trend 500 at Riverside. He finished 14th—ahead of such series hotshoes as Fred Lorenzen, Ned Jarrett, Parnelli Jones, and Dick Hutcherson. *JDC Collection*

When the NASCAR tour rolled into Daytona for 1963, all eyes were on the Chevrolet teams. Word of their new engine's power had leaked to the press, and even before Johnson's Ray Fox-prepped Impala rumbled out onto the track, Fomoco drivers were raising a ruckus about the car's legality. And, of course, they had a point. No one was contending that the new Mystery Motor was a regular production option. But even the "crate motor," over-the-counter exemption that Pontiac 421 engines had enjoyed didn't really provide much homologation cover for an engine line that consisted of fewer than 50 castings.

Ford execs decided to press the point by having John Holman (one half of Ford's Holman & Moody race factory) drop by his local Chevy dealership to order up a Mystery Motor of his own for testing. Caught between a rock and a hard place and hoping at all costs to avoid any publicity that might bring out the legal eagles, Chevy execs actually sold Holman a copy. And that's probably the only Mystery Motor that ever got delivered to someone not named Fox, Yunick, or Johnson.

Junior Johnson proved the validity of Ford racers' concerns by quickly destroying the track lap record at Daytona during testing. His white #3 Impala tripped the clocks at an incredible 165.183 miles per hour during second-round qualifying, a speed nearly 10 miles per hour faster than Fireball Roberts' pole-winning speed of just one year before. Junior backed up his fast lap with 40 more to win the first qualifying race for the 500. Indy ace Johnny Rutherford drove Smokey Yunick's #13 Mystery-Motored Impala to an equally impressive rookie win in the second pre-500 qualifier, and his black-and-gold car was followed closely across the stripe by Rex White's

The number on Nat Reeders' Impala is symbolic of both the amount of factory backing and the number of wins that Chevrolet drivers racked up in 1965 (the season he campaigned the car in the Grand National ranks). It would take five more years for Bow Tie drivers to begin visiting victory lane again with any regularity. *JDC Collection*

#4 MkII 427-motivated Chevy. Rutherford's victory was both the only win ever scored by a NASCAR driver in his first Grand National event and Rutherford's sole NASCAR triumph.

Unfortunately, new engine teething problems hobbled the Mystery Motor drivers during the race proper. Johnson's rocket retired on lap 26 with distributor problems while Rutherford and White finished laps down to the winner Tiny Lund in a Wood Brothers-prepped Ford.

And that's pretty much the way the season played out for the now non-factory-backed Chevy teams—great qualifying speed followed by mechanical gremlins during the race itself. Johnson ultimately recorded superspeedway wins at Atlanta and Charlotte (along with a handful of short-track victories). But by the end of the season, his Holly Farms team was so short on replacement parts that he had to go to Holman & Moody and buy back the spare Mystery Motor that Chevrolet had sold the legendary Ford race shop at the beginning of the season.

Joe Weatherly was another "de-factory-backed" racer who struggled on during the 1963 season. Though he started out in Bud Moore's #8 Catalina, when Moore's funds dried up, Little Joe hopped rides wherever he could find them until Moore signed on with Mercury at season's end. The two Pontiac wins that Weatherly scored (one in the Rebel 300 at Darlington) helped him win a second straight Grand National driving title. Sadly, it was to be the well-liked Weatherly's last, as he was killed in a racing shunt at Riverside Raceway in early 1964.

1964–1969: Independents Carry the Banner

In a strange way, Weatherly's death paralleled the death of GM drivers' Grand National aspirations for the balance of the decade. Though independents like Smokey Yunick and Bobby Allison soldiered on with GM mounts at selected races, not one Oldsmobile, Pontiac, or Buick win was recorded between 1964 and the late 1970s.

Chevrolet drivers managed only a paltry eight wins between 1964 and 1971 themselves. Still, Bow Tie-equipped NASCAR race cars did make some news during the remainder of the 1960s.

Wendell Scott's Chevrolet

Wendell Scott, the series' first black driver, scored his first Grand National win in a 1962 Chevrolet at a 100-lap race in Jacksonville, Florida, in December 1963. The journeyman racer from Danville, Virginia, personified the term "independent driver," and his victory at Jacksonville over the likes of Buck Baker, Jack Smith, Richard Petty, and Joe Weatherly, though not confirmed until after the event due to a scoring mix-up, was still a historic one. As things turned out, it was Scott's only full series victory. His Chevy win was also the

Once upon a time back when the Earth was still cooling, each of the General's divisions relied on engines of their own individual design—both on and off the track. This is what a racing Oldsmobile power plant looked like around 1965. All of that changed in 1978 when the sanctioning body recognized the Chevrolet small-block engine as the corporate engine for all GM race cars. *JDC Collection*

NASCAR stockers had lost a good deal of their "stockness" by the mid-1960s, as can be seen in this shot of a Smokey Yunick-prepared 1966 Chevelle. Then again, it's been said that Smokey's cars were always significantly less "stock" than those of his competitors. Lowered over a handmade frame and riding on handmade suspension components, Smokey's Chevelles were sleek, fast, and probably all "illegal."

only visit paid to victory lane by a GM driver until 1966.

Bobby Allison's Chevelle

Bobby Allison provided the next race victories for the Chevrolet division in 1966 with his home-built 1965 Chevelle. As longtime NASCAR fans will recall, Bobby Allison was one of the founding members of the "Alabama gang" and as independent minded a racer as there was in the mid-1960s. Allison's career had gotten off to a start on the hard scrabble modified tour while he and brothers Donnie and Eddie were still growing up in Miami, Florida.

By the mid-1960s, Allison had moved up to the Grand National ranks and had served as a journeyman driver for a number of teams. Though he'd piloted many different brands of race cars for others, Bobby's personal competition machine at the time had Bow Ties all over it. When not campaigning someone else's car, Allison's mount for 1966 was a diminutive #2 Chevelle that he had built in his two-car garage on Berry Drive in Hueytown, Alabama.

It was nothing short of heresy for Allison to expect to win on the Grand National division with a "Brand X" car. After all, Ford and Chrysler were the big dogs on the tour, and their factory-backed racing efforts were spending money by the bushel basket full, all in an attempt to win on Sunday. It just wasn't possible for an unfunded independent like Allison to presume he had a shot at victory—let alone while he was driving a home-built Chevrolet of all things.

While that may have been the conventional wisdom in those days, it seems that nobody bothered to tell Allison that chances for a Grand National victory were remote. And it probably wouldn't have mattered if they did.

You see, Allison had an ace in the hole. Actually, it was an ace under his hood. And that was the 327 engine that his tiny Chevelle relied on for power. NASCAR rules of the day linked weight to displacement. Specifically, Grand National (GN) cars in 1966 had to weigh 9.36 pounds for every cubic inch they displaced. Factory Fords and Dodges running 427- and 426-cubic-inch engines had to weigh in at just short of 2 tons as a result. Allison's 327-powered Chevelle got to diet down to a racier 3,060.72 pounds, by dint of the "mouse" motor under its hood. And that lighter weight paid big dividends at the track.

Allison's other advantage was the "front steer" suspension setup that Chevelles came factory equipped with. He felt that it offered better "feel" than the "rear steer" system that Ralph Moody had perfected on his Holman & Moody-built Galaxies.

Allison put those ingredients together at Oxford, Maine, in July 1966 to score the first

Chevrolet GN win in nearly three years. After the race Allison said of his lightweight little Chevelle: "I think a new wrinkle has started. We're pleased in two ways. It [the Chevelle] handled well right off the trailer. And secondly, it blew right by all the hot dogs."

Allison blew right by hot dogs like Ned Jarrett, David Pearson, Richard Petty, and Buddy Baker a second time four days later at a 60-mile event at Islip Speedway and yet again in August 1966 at Beltsville in the Maryland 200. Though those short-track wins weren't the headline producers that a Daytona 500 victory might have been, they were cause for Chevrolet fans in the stands to take heart.

When the factories took note of Allison's under-funded performance, they immediately elected to remove the Chevelle-sized thorn in their paw by giving him a full ride in a factory-backed car. First in line was Fomoco, who offered him a seat in Bud Moore's Mercury. When that relationship soured in 1967, Allison dusted off the Chevelle and then dusted off the competition at GN races in Winston-Salem and Savannah. Next came the chance to drive a Cotton Owens Dodge, but Allison's stint as a factory driver came to an end after he won a third race in his Chevelle (at Oxford, Maine) at a race that Owens had elected to sit out. Chrysler execs were less than enthusiastic about a Chevrolet win scored by their "Dodge" driver, and so Allison was an independent once again. That lasted until late 1967 when Allison was offered a full Holman & Moody ride with former champ Fred Lorenzen as team manager. Allison scored one more win in his Chevelle in July 1968 back at Islip Speedway just after he quit his factory ride (again!) that season. It was to be the last GN win scored by a GM driver that decade. But things would get dramatically better once the 1960s were over.

Smokey Yunick's Chevelles

Smokey Yunick also provided some encouraging moments for Chevy fans in the 1960s. As independent a thinker as Allison,

Smokey Yunick's legendary mid-1960s Chevelles all relied on various versions of the Chevrolet big-block engine for power. One of Smokey's speed secrets in those days was to destroke a rat motor to less than the legal 427 cubic inches the rules book allowed in order to secure a few extra rpm down the back stretch.

Smokey believed that if the rules book didn't specifically prohibit a particular piece of mechanical equipment, then it must be legal. And that very liberal interpretation of the sanctioning body's edicts led directly to a string of truly memorable Chevy stock cars.

The years 1965 and 1966 offered independents like Smokey more of a chance at victory as Chrysler and Ford (in turn) boycotted the Grand National tour during these seasons. The two automotive giants took issue with Big Bill France's always-confusing rulings about factory-backed Hemi race engines. Taking advantage of the situation, Smokey cooked up the first of his black-and-gold Chevelle groundbreakers for the August 1966 running of the Dixie 400 at Atlanta.

The absence of big-named Ford Motor Company (Fomoco) drivers and the dominance by factory cars of just one stripe (while the other carmaker sat on the sidelines) had a predictably negative impact on gate receipts.

Without factory backing, Pontiac drivers like Ken Spikes had little real hope of winning races on the NASCAR tour during the mid-1960s. But that didn't stop the cars they campaigned from looking good. *JDC Collection*

Promoters were howling for France's head by the summer of 1966, and he was desperate to do something to create at least the appearance of multi-make competition out on the track. Since Dodge and Plymouth drivers were winning just about everything that year while Fomoco types stewed, word from on high leaked out that tech inspection wouldn't be all that rigorous for any independent Ford or Chevrolet teams that decided to show up at Atlanta. Both Junior Johnson and Smokey heeded the call, and the end result was two cars that will forever live in NASCAR lore.

Johnson's entry was yellow and had, at least at one point in the past, started out as a 1966 Galaxy. In Atlanta trim the car looked anything but stock. The top was chopped, the nose looked squashed, and the bustle rose toward the sky in a most un-stock-like way. Wags around the garage area took to calling the car the "yellow banana" due to its very suspect configuration. But NASCAR tech officials deemed the car legal to race.

Smokey's approach was subtler but just as direct. His efforts to reduce aerodynamic drag focused on making his Chevelle smaller in just about every dimension. According to those who actually got to appreciate his handicraft, the car looked normal until someone actually compared it to an, ahem, full-sized stock Chevelle. Fellow Chevelle racer Bobby Johns for one, took notice of the car's shrunken silhouette, and soon people were calling Smokey's #13 car the 15/16th Chevelle. It, too, was deemed legal to race.

Unfortunately for both Johnson and Yunick, race victory was not in the future for either car. Lorenzen hit the wall hard in Johnson's banana on lap 139 just after Curtis Turner had retired Smokey's car with a broken distributor. Neither car was allowed to run again, and shortly thereafter NASCAR instituted the first body template rules.

Smokey's next headline-producing Chevelle was built for duty in the 1967 Daytona 500. Smokey made a dramatic entry to the garage area on pole qualifying day with his #13 Chevelle shrouded in mystery under a car cover on the trailer. Once unloaded and in its stall, Smokey, driver Turner, and the rest of the crew left the car alone until the last possible moment for a shot at the pole. Turner then climbed in, fired the car up, and took it out on the track for its run at the pole. Two laps later he'd set a new lap record of 180.831 miles per hour— more than 5 miles per hour faster than Richard Petty's 1966 pole-winning speed. It was an incredible performance, to say the least, and factory teams were furious. To say that Smokey had stolen their thunder was an understatement. And

Following pages
Rex White drove a "bubble-top" Chevrolet much like this one during the 1961 season. Though a much better looking machine than it's GM and Brand X rivals, it unfortunately was down on power when compared to them. Florida's Harold Doherty built this replica of White's Impala. *JDC Collection*

Though Olds drivers dominated the first few years of the NASCAR series in the late 1940s and early 1950s, by the mid-1960s, the few hearty souls who ventured out onto a Grand National track under Oldsmobile power had little chance of victory. Had drivers like this unknown 1969 independent enjoyed the same factory backing that their Fomoco and Chrysler rivals did, they probably could have invested in the components (not to mention trailers) necessary to win. *JDC Collection*

DAYTONA
REX WHITE
NALLEY CHEVROLET, INC.
NALLEY
NICHOLSON
DYNO SHOP
ATLANTA,
GA.
4
PPG
Pedrick
PISTON
RINGS

AIR LIFT
EQUIPPED WITH
CHAMPION
SPARK PLUGS
te & Clements
Garage
tanburg, S.C.
CHEVROLET

yet, to hear Smokey tell the story today, his only speed secret was destroking the displacement of his Chevelle's 427 big block down to 410 cubic inches.

Whatever the basis for the car's speed was, Turner was only able to make 25 circuits of the track during the race before the engine blew. Smokey rebuilt the motor and turned up with Turner at Atlanta in April. The #13 Chevelle was once again the fastest thing at the track during testing. That is until Turner demolished the car in practice in the most dramatically airborne way possible. Following the shunt, nervous Chevy execs who had been slipping Smokey parts on the QT tried to "repossess" the remains of the Chevelle. So Smokey sent the car back to Michigan. But not until he'd crushed the car into a cube. It doesn't pay to make Smokey Yunick mad.

The third and final Chevelle that Smokey created was built for duty in the 1968 Daytona 500. Like the others, it started life as a 1966 model, but by the time it rolled off the truck in the garage area, it was anything but. Smokey was an early student and advocate of aerodynamics, and he also knew how important optimum weight distribution was for on-track handling. His last little Chevelle was built with an eye toward both of those variables.

Construction of the car commenced with the fabrication of what was probably the first purpose-built tube frame chassis in NASCAR history. Smokey then installed the car's body over the car as far to the rear as possible for improved weight distribution. He also biased the body as far to the left as possible for the same reason. The car's floorboard was lowered to both serve as an aerodynamic belly pan and to keep planned driver Gordon Johncock's weight as low as possible in the car. Sectioning the front bumper and adding an extra inch and a half strip of metal before re-chroming further enhanced aerodynamics and handling. The end result was a rudimentary front airfoil that worked in concert with a vortex-generating channel in the roof panel to keep the car glued to the ground at speed.

Glenn "Fireball" Roberts was arguably the first superstar in the NASCAR series. He was college educated, articulate, and photogenic. He also happened to go fast as stink while at the wheel of a stock car (such as the Smokey Yunick-built Poncho he won the 1962 Daytona 500 with). *JDC Collection*

Unfortunately, when Smokey tried to get the car through tech at Daytona, he was handed a long list of things he'd have to change to make it legal to run. First on that list was "remove frame and replace with stock." Smokey responded to that list by firing the car up and driving it through traffic across town to his Daytona Beach garage. Bill France announced to the press the next day that Smokey had made that five-mile trip without the benefit of having a gas tank in the car. The car was never allowed to turn even one lap on a NASCAR track, but it still became a racing legend.

The 1960s, though brimful with promise at the dawn of the decade, turned out to be an unmitigated failure for the "General's" stock car racing efforts. When the 1960s ended, few GM fans would have guessed that better days for Chevrolet drivers were just around the corner. But indeed they were.

16
366-C.I.
DieHard
AMC Matador
Matador
GOODYEAR
14
427 C.I.
Fitzgerald
CUNNINGHAM-KELLEY
CHEVROLET
GOODYEAR
union

Chapter

THE 1970s

THE BOW TIE RETURNS

Though Fomoco and Chrysler cars had dominated the Grand National series during the 1960s, by 1970 their collective interest in motorsports was on the wane. The federal government had begun making increasing demands on the major automotive manufacturers in the areas of air pollution and occupant safety. That fact, coupled with the escalating price of fossil fuels, caused executives like Lee Iacocca to reconsider their corporations' commitment to factory-backed auto racing. Iacocca cut Ford's racing budget by 75 percent upon his ascension to Ford's top spot in late 1969, for example, and by 1970 Ford was out of racing altogether. Chrysler continued to fund factory-backed teams for just one more season, and then they too pulled up stakes and went home.

Sterling Marlin is a second-generation NASCAR driver. His father, Coo Coo Marlin, was a journeyman driver and Chevrolet partisan during the 1970s. The senior Marlin is pictured here in his 1970 Monte Carlo. The highlight of Coo Coo's career was his surprise win in one of the twin 125 qualifiers for the 1973 Daytona 500 in the Monte Carlo pictured above. *JDC Collection*

The sudden departure of factory funding left both the sanctioning body and racers in disarray. Winston would ultimately step in to fund the series in 1972, but the loss of factory R&D and subsidized high-performance parts made it tough for teams to field competitive cars. Those teams (like Petty Engineering) that had stockpiled the largest inventory of racing exotica initially fared the best, but ultimately the supply of those high-dollar, race-only components began to wane.

Luckily for racers like Junior Johnson, though Chevrolet had been on the racing sidelines during the balance of the 1960s, the corporation had never ceased the production of high-performance engine and drivetrain parts. Better yet was the plentiful and economical nature of those parts. You see, unlike Chrysler and Ford, who had focused on exotic engines like the Boss 429 and the 426 Hemi Chevrolet's focus on more common engines like the 427 "rat" motor and the 327-350 "mouse" motor not only created an abundant supply of factory-cast parts, it also resulted in a huge support network of aftermarket suppliers who were already making Chevy go-fast goodies for the legions of Bow Tie fans and their street cars.

Cale Yarborough enjoyed great success as a Chevrolet driver during the mid-1970s. One of the secrets of that success was the drooped snout on his Laguna S-3 Chevelle race cars. *Mike Slade*

1971–1972: Junior Johnson Leads the Comeback

As a result, when Ford factory support for Junior Johnson's Torino Talladega team dried up in early 1971, it was a no-brainer for him to return to his General Motors roots and switch back to a Chevrolet-based team. Johnson combined forces with Charlotte Motor Speedway's Richard Howard and placed an order with Chevrolet for a handful of its new-for-1970 Monte Carlo bodies and a sufficient number of 427 crate motors to last a full season. Key personnel on the new team included driver "Chargin'" Charlie Glotzbach, legendary crew chief Herb Nab, and a young Holman & Moody line mechanic named Robert Yates.

It took several months out of the first part of the 1971 season for Johnson to get his new team up and running. In race trim his new Monte Carlos sported the same white #3 racing livery that Junior's Mystery Motor Impalas had in 1963, right down to the red-and-blue stripes that ran from hood to trunk lid. Race number one for the new Chevy team took place at Charlotte in the Memorial Day running of the World 600. Charlie Glotzbach made sure that everyone knew of the team's arrival on the Grand National scene by ripping off a pole-winning lap of 157.788 miles per hour during qualifying.

During the race itself Glotzbach battled with Bobby Allison's Holman & Moody Mercury for the lead until a late-race crash removed the Monte Carlo from contention. Win number one for the team was not long coming, however, and was scored at Bristol in the Volunteer 500 in July 1971. Though that triumph turned out to be the team's only win of the season, it signaled the end of Chevrolet's long drought on the NASCAR circuit.

In 1972 the team added Bobby Allison as team driver and the Coca-Cola sponsorship package that he brought with him. The new combination was successful right out of the box. Win number one for Allison's "Coke Machine" came at Daytona in the second of the traditional Twin 125 qualifiers that precede

Following pages
Cale Yarborough's Olds Cutlass was quite the aero-warrior. Though its Busch livery was arguably not as fetching as the Holly Farms Chevelles Yarborough had campaigned before, there's no doubt that it was just as fast. *Mike Slade*

Junior Johnson supplied Holly Farms-backed (among other sponsors) Chevelles and Monte Carlos for South Carolina native Cale Yarborough during the 1970s. Yarborough returned the favor by winning the Winston Cup championship three times in a row for Johnson's team. *Mike Slade*

aceway
BUSCH
11
Cale Yarborough
NASCAR
RACE CAR
CHAMPION
STP
Edelbrock
MANIFOLDS
Holley
CARBS
76
GASOLINE
RAM
MOROSO

BUSCH
DieHard
BUSCH
HA
SPEED PRO
MOOG
VALVOLINE
11
11

Though the big-fendered 1973–1977 Monte Carlo line might at first glance seem to be aerodynamically challenged, it actually turned out to be a pretty stout comp car. This particular car, for example, played an integral part in Dale Earnhardt's first Winston Cup championship.

Earnhardt's front office looked something like this in 1980. Note the Dodge A-100 van seat that Ironhead still prefers today. Try to find a stock component somewhere in this photo.

the 500. The team's first win in a full-scale superspeedway event came one month later when Allison edged A. J. Foyt's Wood Brothers Mercury in the Atlanta 500. When Allison won again at Bristol in the Southeastern 500, Chrysler and Ford teams began to grouse about the rules book's preference for wedge motors like his 427 (1972-era restricter plates were more generously proportioned for wedge motors than for Hemis). It was a sure sign that the Coke team was hitting them where it hurt. Allison kept up that pummeling through to the end of the season.

All told, Allison and his #12 Monte Carlo scored 10 wins in 1972, including a stop at victory lane in the all-important Southern 500 at Darlington. Those wins and 15 other top-five finishes earned Allison a second-place berth in the seasonal points chase and $348,405 in winnings. The results of the first season in what is now called NASCAR's "modern era" foretold Chevrolet fortunes in the balance of the decade. Chevrolet was back—in a very big way.

1973–1974: Loading up the Chevy Bandwagon

It didn't take long for other Grand National teams to sit up and take notice of the success enjoyed by Junior Johnson's Chevrolet-based team. As supplies of Chrysler and Fomoco exotica began to disappear, more and more teams made the switch to GM. Since Johnson had been there "firstest with the mostest," it will come as no surprise that his Chevy-based team continued to enjoy the most success with Bow Tie-based comp cars. Cale Yarborough signed on as team driver in 1973, and for the next six seasons his Junior Johnson-equipped GM stockers were the class of every racing grid they graced.

Adding to Chevy's success in 1973 was Benny Parsons who earned the first Grand National (now Winston Cup) driving championship of the modern era at the helm of L. G. DeWitt's Monte Carlos and Chevelles. And he

Earnhardt's first quality ride on the Winston Cup tour was provided by Rod Osterlund in 1980. Short-track duty was handled by big-fendered Monte Carlos like this one currently on display at the International Motorsports Hall of Fame in Talladega.

did it the old fashioned way—by running consistently in every race he entered. In fact, Parsons scored just one race win (at Bristol in the Volunteer 500) during the entire season, but though not often in first place, he wasn't far from it. By season's end Parsons had notched 15 top-five finishes and 21 top tens on his way to the title and $128,321 in winnings.

The last gasp for Chrysler drivers on the Winston Cup tour came in 1974. Richard Petty won his final non-GM national title that year, but the Chevrolet writing was already on the wall of his Randleman shop by the end of the season. And much of it was written by "graffitists" Junior Johnson and Cale Yarborough. Cale's #11 Carling Chevelle graced the winner's circle in Darlington at the Southern 500 that year as well as at nine other tracks.

The venerable Chevrolet small block has been winning races since its introduction in 1955. Dale Earnhardt relied heavily on his small blocks to provide the "motorvation" for his first Winston Cup championship in 1980.

STP
43
GOODYEAR
GOODYEAR
REED
CHAMPION
STP
DieHard
UNION 76
PEAK
76
GASOLINE
Holley
LF

Richard Petty jumped ship for General Motors in late 1978. His trademark red-and-blue colors first showed up (in the modern era) on Chevy sheet metal at Michigan in August of that year. *Mike Slade*

Chevrolet wins for the first four years of the new decade totaled 32, and the future was looking bright. But truly the best was yet to come.

1975: Small-Blocks Invade the Scene

By 1975 the big-block NASCAR race engine had become a thing of the past. And that included the by-then venerable Chevrolet "rat" motor. First introduced in Mystery Motor form in 1963, the Chevy poly-angle-valved, big-block wedge motor was never as exotic as, say, a Boss 429, 426 Hemi, or 427 single overhead cam (SOHC) motor, but on balance, it was probably a more effective overall performer. Though deprived of racing glory during the 1960s when it was first homologated in 396 form, Chevy's rat at least got a brief moment in victory lane during the first few years of the modern era. Interestingly, its competition replacement was an engine even older than the Mystery Motor. And that, of course, was the incredible Chevrolet small-block race engine, which by 1974 had evolved into a full 358 cubic inches.

Cale Yarborough and Junior Johnson would erase all doubts about the small-block Chevrolet engine's potential as a championship contender during the 1976, 1977, and 1978 Winston Cup seasons. By the mid-1970s the 350-cubic-inch "mouse" motor had been massaged to produce more than 450 horsepower. When coupled with the slope-nosed shape of a slippery Chevelle Laguna S-3 body, speeds in excess of 200 miles per hour were just a stab of the throttle away. And that potential was quickly translated into superspeedway wins by Yarborough and the Holly Farms team.

Darrell Waltrip used the slippery shape of this Oldsmobile Cutlass to score a big superspeedway win in the 1979 running of the Talladega 500. The "corporate" Chevrolet small-block engine under the Gatorade car's sloping hood probably played a role in that victory too. *JDC Collection*

1976–1977: Championship Seasons for Chevy and Yarborough

When not rocketing around the high banks, Johnson's team relied on boxier Monte Carlo chassis for short-track duty. Though not nearly as sleek as the Laguna S-3, a 1973–1977-style Monte Carlo's more generously proportioned flanks allowed for plenty of fender rubbing on the shorter bull-ring circuits. The car's long snout and short trunk also provided better weight distribution (due to engine setback), which in turn enhanced handling. Yarborough and company relied on both body styles to rack up 9 wins and 22 top-five finishes on the way to his first Winston Cup championship in 1976. Chevrolet wins totaled 13 that year and grew to 21 the following season.

Yarborough and his swoopy S-3 were back at Daytona for 1977, though probably not greeted with much enthusiasm by the other teams in the garage area. He used what he'd learned during his win in the 1976 Firecracker (also held at Daytona) to win his second Daytona 500 that year. He and Johnson then steamrolled the competition in 8 more events, winning 2 more superspeedway races and 6 out of the 10 short-track events he entered. By season's end he'd secured his second Winston Cup championship.

Yarborough's only real competition that season was a brash young Kentuckian named Darrell Waltrip, who campaigned Chevrolets for the Di-Gard team. Collectively, Waltrip and Yarborough won 15 of the 30 races that made up the 1977 season and finished in the top five at 26 others. Along the way, Waltrip's knack for self-promotion inspired Yarborough to saddle him with the nickname "Jaws" and

Like most Chevrolet teams on the tour, Dale Earnhardt and Rod Osterlund elected to campaign the Olds Cutlass chassis at superspeedway events during the late 1970s and early 1980s. The car's slippery shape coupled with tried-and-true small-block Chevy power helped Earnhardt win his first Winston Cup crown in 1980.

The small-block Chevrolet has become a familiar sight in the Winston Cup garage area. Dale Earnhardt relied on a "mouse" motor for every one of his seven Winston Cup titles. Though limited to just 358 cubic inches, a modern small-block motor can churn out more (700-plus) horsepower than one of its big-block brothers from the early 1970s.

North Carolina's Alex Beam has recently restored one of Dale Earnhardt's championship-winning 1977 Olds Cutlasses. Its angled snout and fastback roofline made the car plenty quick on the big tracks.

the rivalry (if not the affection) between them only intensified.

1978: Oldsmobile Takes the Baton

The ever-changing NASCAR rules book struck Chevrolet fans a blow during the 1978 season. The sanctioning body decided during the off-season that the Laguna S-3's drooping snout gave Chevy drivers a bit too much of an aerodynamic edge over their boxier competition and so, with a wave of the rules-book pen, banned the body style from competition.

Fortunately for GM partisans, sister division Oldsmobile had already built a 442 Cutlass body style (in 1977) that featured an angled beak that was nearly as aerodynamically efficient as the S-3's snout, and the NASCAR rules book said nothing about banning it. Best of all, the sanctioning body decided to recognize the

Richard Petty jumped the Chrysler ship for General Motors in late 1978. It proved to be a wise move when, in 1979, Petty and a fleet of Monte Carlos and Cutlasses won the King's seventh Winston Cup title. Petty drove this car to victory in the 1979 Daytona 500. *Mike Slade*

Mid-seventies Monte Carlos were boxy and seemingly ungainly. Nonetheless, they made superb race cars. Their long hood lines and short rear deck allowed big time engine rear set, for example, and that helped weight distribution a bunch. *Mike Slade*

Chevrolet small block as a "corporate" engine that same year. That meant that Chevrolet racers interested in swapping to Olds sheet metal in search of an aerodynamic edge could still rely on their race proven "mouse" motors.

So that's just what Cale Yarborough and Junior Johnson did for the 1978 season. Cale scored the Oldsmobile division's first NASCAR win in 18 years in January of that year when he came home first in the Winston Western 500 at Riverside. In May he put his new Olds' slippery shape to good use by winning the Winston 500 at Talladega from the pole (191.904-mile-per-hour pole speed). Big-track wins at Michigan in June and Darlington in the Southern 500 followed in short order. Those wins, coupled with 6 other visits to victory lane and 23 top-five finishes, added up to Yarborough's third straight Winston Cup title.

Olds wins that year came in at 11, while Chevrolet victories (scored mostly on short tracks where generously fendered Monte Carlos still ruled) fell to 10.

The other big news in the GM ranks for 1978 was Richard Petty's decision to jump ship from Chrysler Company cars and return to the GM fold. King Richard made the change late in the season at Michigan, where he debuted a #43 Monte Carlo at the August 400-mile race. Though Petty's mechanical change of allegiance did not bear fruit in 1978, things came up roses for him the following season.

1979: The King and His Cutlass Reign

As Petty adherents are sure to recall, the King's first competitive outings in the NASCAR ranks came at the wheel of one of his dad's Oldsmobile stockers. And it was in a #43 Olds Cutlass that Petty snapped a 45-race winless streak by winning the 1979 Daytona 500. Petty's win was made all the more spectacular by the fact that he'd had a large portion of his stomach removed during the off-season (due to ulcers) and was supposed to be sitting at home recuperating instead of averaging 143 miles per hour on his way to a sixth win in the 500.

Petty followed up on that auspicious start to the season by adding four more wins to his even then-impressive total. Twenty-three top-five finishes gave Petty an unprecedented seventh national driving title. Olds wins totaled five for 1979, and all came at superspeedway events. Petty and most other GM drivers continued to rely on their bulbous-fendered Monte Carlos for the bulk of their on-track time in 1979. And those unlovely, if effective, cars visited the winner's circle 18 times as a result.

Dale Earnhardt was a first-time winner on the Winston Cup tour that year in his #2 Monte Carlo at Bristol in the Southeastern 500, and Darrell Waltrip added seven Chevy wins to his

Chevrolet Monte Carlos (like Bobby Allison's "Coke Machine" pictured here) were winners right out of the box. In fact, the car line has racked up more NASCAR wins than any other GM body style. *JDC Collection*

The secret of the Laguna S-3's success was its sloped front facia. It cut through the car much better than cars that carried more upright grilles. And that meant faster speeds on the track. *JDC Collection*

tally with the help of a #88 Monte Carlo that same season. Bigger things were just around the corner for Waltrip as 1979 drew to a close.

It is clear now that the 1970s truly belonged to Chevrolet and General Motors in NASCAR competition. Though few would have predicted it when the decade dawned, Chevrolet teams scored 90 Winston Cup wins between 1971 and 1980 and added 5 Winston Cup titles to that total for good measure. Though the 1970s were kind to GM drivers, the best was truly yet to come.

Though boxy and squared off, Bobby Allison's 1971 Monte Carlo was still plenty fast on the super speedways. It's likely the full Race Rat motor under the car's hood had a little something to do with that. *Daytona Racing Archives*

11
11
DieHard

Chapter

THE 1980s

COMPETITION HEIGHTENS

With Ford forces reduced to combing junkyards for "racing" gear, the 1980s were destined to be the golden decade for GM drivers.

The cars became faster and smaller with improvements throughout the decade. New models Buick Regal, Monte Carlo SS, and Lumina took center stage when aerodynamic improvements, creating a second Aero War in 1986 and 1987, changed the role of aerodynamics in the racing world.

The era also featured GM cars taking the Winston Cup for 9 of the 10 years. Legendary drivers such as Dale Earnhardt, Darrell Waltrip, Bobby Allison, Rusty Wallace, and Terry Labonte made such a feat possible. Allison, Wallace, and Waltrip won their titles only during this decade.

1980: Last Year for "Big" Cars

The 1980 season was to be the last for cars rolling on 115-inch wheelbased chassis. Grand National and Winston Cup cars had rolled into battle over chassis of that length for what seemed like forever. But that all had to change for 1981 as the days of the "full-sized" car had come and gone on the NASCAR tour. The days of the block-long Galaxies, Impalas, and Marauders that had originally set the 115-inch wheelbase limit were long since over, and after that season so too would be descendants of those cars like the mid-1970s-style Monte Carlos, Cutlasses, and Thunderbirds.

But before those lumbering throwbacks were relegated to the boneyard, they were allowed to have one last season in the sun. Buddy Baker kicked off that last year of the behemoths with an Olds Cutlass win in the Daytona 500. Darrell Waltrip and Junior Johnson continued their winning ways for their respective Chevrolet teams, and newcomers like "Texas" Terry Labonte scored first wins in their Chevrolets. But the year belonged to an up-and-coming second-generation stock car driver named Dale Earnhardt.

As is well known now, "Ironhead" is modified champ Ralph Earnhardt's son. And in the Earnhardt family, blood does really tell when it comes to driving skill. Dale Earnhardt's first stab at the Winston Cup ranks had come back

The colorful Buicks that Darrell Waltrip used to pummel the competition during the 1981 and 1982 seasons owed much of their speed to the drooping hoodline and angled grille that were part of the Regal styling package those years. *JDC Collection*

Waltrip's Buicks looked good from just about any angle—except perhaps from the rear. Unfortunately, that's about the only view that D.W.'s competition got to check out during the 1981 and 1982 Winston Cup seasons.

in 1975, and for the next few seasons he skipped around the series in search of a quality ride. By 1979 Earnhardt was a seasoned driver and had hooked up with Rod Osterlund's Chevrolet-based team. It proved to be a winning combination. Earnhardt accomplished the nearly impossible by winning a race in his first full year on the tour and taking Rookie of the Year honors. The very next season (1980), however, he managed to do the completely impossible by winning the Winston Cup championship in only his second full year on the tour. A big part of that title, of course, can be attributed to the yellow-and-blue #2 Chevrolet Monte Carlo that Osterland set up for Earnhardt every week. He recorded 5 wins that season and finished in the top five in 14 other races on his way to Winston Cup title number one. It wasn't to be his last.

1981–1982: Buicks Return to Victory Lane

As mentioned, 1981 was the first year for NASCAR's new downsizing rule. Overnight, the 115-inch wheelbase cars that had been the standard for nearly two decades became out of date, and teams were forced to build (or cut down) chassis based on a 110-inch wheelbase. All bets were off for every manufacturer, and each team had to essentially start fresh in learning how to make the new, shorter cars handle.

The first team to learn how to do just that belonged to—not surprisingly—wily old veteran Junior Johnson. Never one to be afraid to try something new if it seemed to hold mechanical promise, Johnson was one of the first to abandon the Monte Carlo line in favor of the Buick Regal body style, the reason being aerodynamics. Unlike its Chevrolet and Pontiac siblings, the Buick Regal featured a forward-sloping grille and tapered hood line that was very efficient at cutting through the air at superspeedway velocities.

Power for Johnson's Mountain Dew-backed team that season was provided by a pumped-up 358-cubic-inch small block. The soda maker's green-and-white racing livery graced the #11 car, making it one of the best-looking Cup cars on any starting grid. Darrell Waltrip's driving skill also made it one of the fastest.

Waltrip slipped behind the wheel of Johnson's Dew-mobile for the season opener at Riverside after having to buy himself out of his old DiGard contract. It turned out to be a bargain at twice the price. Win number one for the new combination came at Rockingham in March, and from there on, Waltrip never looked back. By season's end, the flashy green-and-white Buick had won 12 races and notched 21 top-five finishes. The 4,880 points Waltrip accumulated were enough for him to win his first Winston Cup crown.

Fellow Buick drivers Richard Petty, Bobby Allison, and Cale Yarborough did their part to bring Buick's win total to a whopping 22, and those were the first marks in Buick's win column since Fireball Roberts' victories way back in 1955. Morgan Shepherd's win at Martinsville in the Virginia 500 was the first scored by a Pontiac driver in nearly as long (since 1963). Chevy drivers did not fare well, recording only a single win in the first year of the 110-inch wheelbase rule.

The Dew Crew had another big year on the Winston Cup tour in 1982. But they didn't begin their 1982 "tear" until after Bobby Allison had scored a popular victory in the Daytona 500 with his #88 DiGard Buick. It was the first "Super Bowl" win for that manufacturer and foretold the year that Buick drivers would enjoy on the tour. Waltrip's win total for 1982 came in again at 12. And as in 1981, fellow Buick driver Bobby Allison came in second behind the Dew car in the seasonal points race.

All told, Buick drivers scored 25 Winston Cup wins. Pontiac drivers were shut out that season, but Monte Carlo drivers managed 4 wins.

That's Morgan Shepherd's #98 Buick getting sandwiched between Bobby Allison's Regal and Richard Petty's Grand Prix. The scene was a typical one on the superspeedways of the early 1980s: no Fords in sight. *JDC Collection*

1983: GM Divisions Divvy Wins

Cale Yarborough began 1983 on an auspicious note for Bow Tie fans by winning the Daytona 500 with a new Monte Carlo SS. Tired of taking a back seat to the Buick Regal in aerodynamic prowess, Chevrolet engineers cooked up a bolt-on beak for the 1983 Monte Carlo line that replaced the car's former upright grille. The new nose featured a flexible fascia that offered a more angled attack to the wall of wind encountered at 200 miles per hour. Allison stood his ground with his Buick and scored win number one for 1983 at Richmond. Richard Petty also returned to the winner's circle in his Pontiac at Rockingham.

Overall the new-and-improved Monte Carlo proved to be the best car on the tour in 1983, winning 15 races. Buick pilots like Allison came in second with 6 wins, and Pontiac drivers scored 5.

Coming out on top of the points race was Bobby Allison, who finally stepped out of the Dew Crew's shadow to crown his spectacular career with a championship finish. Allison accomplished the victory with 6 wins and 18 top-five finishes. Ironically, Darrell Waltrip was second to Allison for a change in his Chevrolet Pepsi Challenger.

1984: Labonte Brings Home the Bacon

Cale Yarborough broke the double ton during qualifying for the 1984 Daytona 500 in his #28 Chevrolet. With a qualifying speed of 201.848 miles per hour, it was the first time that a driver had officially qualified for a Winston Cup race with a speed equal to or in excess of 200 miles per hour. It was an encouraging performance in the face of the aerodynamic prowess displayed by the newly returned to racing Ford Thunderbird line. After sitting on the sidelines since 1971, the Ford Motor Company had decided to return to racing (at least on a limited basis—the days of

Following pages
Though the late 1980s Buick Regal wasn't as aesthetically pleasing as it had been in the early 1980s, it was still a formidable competitor when decked out in NASCAR trim. *Mike Slade*

CHARLOTTE
BUICK
12
76
OFFICIAL FUEL
NASCAR
CRANE
Cams
MAC
QUALITY TOOLS
Stewart
MOROSO
STP
Edelbrock

UNOCAL
BUICK
Miller
12
Winston
UNOCAL
UNOCAL
RICHMOND
GEAR
LOCTITE
FEL-PRO
UNOCAL
RICHMOND
GEAR
LOCTITE
FEL-PRO

All four of the General's performance-oriented divisions were well represented in the NASCAR ranks during the mid-1980s. They collectively dominated the series. Unfortunately, today only Chevrolet and Pontiac are still in the game.

unlimited factory racing budgets are never to return), and for the first time since the mid-1970s, GM drivers had to seriously contend with someone other than fellow GM drivers for wins on the tour.

That having been said, the revitalized forces at Ford still weren't enough to prevent "Texas" Terry Labonte from scoring his first Winston Cup crown in a Billy Hagen–prepared Monte Carlo. Two wins and 17 top-five finishes were enough for the title in 1984. Darrell Waltrip was the most frequent visitor to victory lane in his Budweiser-backed Monte Carlo (7 times), but that total wasn't enough to cinch his third national title. That would come soon enough.

"Texas" Terry Labonte was an Olds pilot in the late 1980s. Though his Piedmont-backed Cutlass enjoyed jet sponsorship, the boxy little car's configuration was less than aerodynamic. Comparing Labonte's car with Richard Petty's swoopy Grand Prix 2+2 will dispel any doubts about that. *Mike Slade*

Harry Gant scored the last NASCAR wins for the Oldsmobile division (to date) during the 1992 season in a #33 Cutlass. The very next season, Olds and Buick decided to drop their backing of the Winston Cup series. One of Gant's Cutlasses has been returned to race duty on the vintage race circuit.

Richard Petty's Pontiac was arguably the prettiest of the long-nosed 2+2s that ran during the 1987 and 1988 seasons. Unfortunately, it never found its way to Victory Lane, but boy it sure looked fast!

1985: More Glory for GM

As you might have guessed, 1985 was another standout year for Chevrolet drivers on the tour. Cale Yarborough notched his second straight Daytona 500 win in his #28 Monte Carlo, and Darrell Waltrip visited victory lane for the first time that year at Bristol. Yarborough took the first Talladega race to under-

Pontiac's answer to the aero-Thunderbird threat during the late 1980s was the Pontiac Grand Prix 2+2. Richard Petty's 2+2 is pictured here. Interestingly, though the cars were built for optimum superspeedway velocity, Rusty Wallace scored the only 2+2 wins on road courses.

The droopy nose and special front fascia of the Grand Prix 2+2 looked fast sitting still. This particular Grand Prix was last raced by Indy ace Al Unser Jr. during 1987–1988 at the now defunct Riverside raceway in California. It is still cutting hot laps today on the vintage race circuit.

score the top-speed potential of his slope-nosed Chevy, and Bobby Allison proved that Buick Regals still had what it took for super-speedway wins by besting all comers in the grueling World 600. In addition, Richard Petty scored career wins 199 and 200 that season aboard a Pontiac. The latter triumph came in storybook fashion at the Firecracker 400 as President Ronald Reagan looked on. His record will, of course, never be topped.

Dale Earnhardt had returned to Richard Childress' team for 1984 after two years as a Ford driver, and his big-track wins (two wins that season and four in 1985) gave a hint of the success the team would enjoy in the near future. But it was Waltrip who stood on top of

No matter what the sheet metal might have suggested, Pontiac racers in the mid- to late 1980s (as today) all relied on "corporate" Chevrolet small-block engines for motivation.

DAYTO
Wrangler
JEANS
Goodwrench
PERFORMANCE PARTS
COMPETITION CAMS
REED
GOODYEAR
EAGLE
tilton
RICHMOND
GEAR

NA • US
Winston Cup
UNOCAL
BUSCH
CHAMPION
True Value
Master MECHANIC
GOODYEAR
Edelbrock
Goody's
STEWART WARNER
TRW
Gatorade
MICHIGAN ENGINE BEARINGS
INGERSOL RAND
STP
JE
SIMPSON
Holley
MOROSO
BILSTEIN
EAGLE
MOOG
Stewart
PERFECT CIRCLE
76
PEAK
EARL'S
GOODYEAR
Stant

the NASCAR hill at the end of 1985, securing his third Winston Cup crown (and car owner Junior Johnson's sixth).

Together, Buick, Pontiac, and Olds drivers went 29 and 0 in 1985.

1986–1987: Aero Wars

Speeds continued to rise at Daytona and Talladega during the mid-1980s. Much of that increased velocity could be traced to the efforts of Chevrolet aerodynamacists. When Ford's sleek Thunderbirds began to get uncomfortably close, GM engineers elected to smooth out the Monte Carlo's formal roofline by grafting on a "bubble-back" rear window and abbreviated trunk lid. The end result was a fastback roofline that more efficiently directed air toward the rear deck spoiler. The new car was dubbed the Monte Carlo SS Aero Coupe, and it was destined to be both one of the best looking and most successful Chevrolet race cars of all time.

Bill Elliott created quite a stir by rocketing around the Big D at more than 205 miles per

One of the reasons for the Monte Carlo Aero Coupe's high-speed prowess was the grafted-on bubble-back rear window that engineers cooked up to create a fastback roof line. It worked.

Previous page
The 1987 and 1988 seasons were the high point of the modern-day factory-backed aero wars. Speeds those seasons regularly topped the 200-mile-per-hour mark. Chevrolet's entry in that competition was the Monte Carlo Aero Coupe—and aerodynamic it was.

Dale Earnhardt became the man in black in 1988 when the Childress team picked up Goodwrench sponsorship. Earnhardt's been bad in black ever since. One of his team cars from the 1988 season is on display at Richard Childress' team museum in Welcome, North Carolina.

hour during pole qualifying for the 1986 Daytona 500. That impressive speed notwithstanding, it was Geoff Bodine's #5 Monte Carlo SS that graced the winner's circle at race end. It was the first of 18 such post-race trips taken by Bow Tie drivers that year. Tim Richmond was the most frequent visitor to victory lane that year; his Chevy victories included road course (Riverside), superspeedway (Darlington Southern 500), and short-track (Richmond) triumphs.

While the always-flashy Richmond was living large in victory lane, Dale Earnhardt was methodically working his way toward a second Winston Cup Championship. His 5 victories that season included triumphs at Atlanta, Charlotte, and Darlington. Sixteen top-5 and 23 top-10 finishes earned Earnhardt 288 more points than fellow Chevrolet driver Darrell Waltrip, and won him the 1986 NASCAR crown.

Earnhardt's second championship season was the same year that Pontiac, Olds, and Buick drivers all returned to victory lane. Terry Labonte scored the first win for a downsized Olds Cutlass at Rockingham in the Goodwrench 500. Pontiac's all-new "bubble-backed" aero warrior, the Grand Prix 2+2, got its first trip to victory lane at Bristol with the help of Rusty Wallace, and Bobby Allison used his #22 Buick to take the Winston 500 at Talladega. As in years past, each of the non-Chevrolet GM victories was achieved with the grunt of corporate Chevrolet small-block, 358-cubic-inch engines.

The high point of the modern era's factory-backed aero wars came in 1987. When GM and Fomoco drivers filed into Daytona for 500 testing that year, all knew that really big speeds would be in the offing. Monte Carlo Aero Coupe and Grand Prix 2+2 teams had had a whole season to refine the performance of their special aero-variant race cars, and Ford troops came packing an all-new and even swoopier version of the Thunderbird.

Ford drivers came out on top in the first test of aerodynamic superiority that season by winning the Daytona 500. Bill Elliott lived up to his awesome appellation at Daytona with a pole-winning lap of 210.364 miles per hour and

The control cabin of Waltrip's Aero Coupe featured an all-encompassing roll cage, a single form-fitting alloy bucket seat, and a brace of analog gauges to monitor engine function.

Darrell Waltrip's last Junior Johnson-built race car was a Budweiser-backed Monte Carlo Aero Coupe. Alex Beam of Davidson, North Carolina, has restored one of those cars to race trim. D.W.'s teammate, the late Neil Bonnett, drove a nearly identical #12 Budweiser Monte Carlo. Between them, they scored 4 wins and 27 top-5 finishes.

a convincing win in the 500 proper. Talladega was also a Ford tour de force as Elliott once again set fast lap time with a blistering 212.809-mile-per-hour circuit, and Davey Allison won the event from a third-place starting position.

Ford's prowess on the superspeedways that season led directly to the reintroduction of the carburetor restrictor plate for the second time in NASCAR history. And those flow-restricting, horsepower-robbing "safety" components are still bunching up cars and causing spectacular wrecks to this very day.

As fast as Ford drivers were on the big tracks, their wins there weren't enough to prevent Dale Earnhardt from scoring his third straight Winston Cup title. His 11 wins bettered Elliott's total by 5, and those triumphs coupled with an incredible 21 top-5 finishes added up to Winston Cup crown number 3 for both Earnhardt and his Monte Carlo.

Pontiac drivers recorded 1987 wins at Riverside and Watkins Glen, while Bobby Allison won the Firecracker 400 in a downsized Buick. Olds drivers did not fare so well, scoring no wins that year.

1988: An Off Year

In many ways, 1988 was the best season and the worst season for longtime GM driver Bobby Allison. Without a doubt one of the season's highlights came in the Daytona 500 when Bobby edged out his son Davey's Thunderbird to claim his second Daytona 500 victory. But that win turned out to be Bobby's last on the tour, due to the near- fatal shunt he would suffer at Pocono in June. Today Bobby is tied with Darrell Waltrip for third on the all-time NASCAR wins list at 84 victories (though his total should be 85 thanks to a Grand American victory that Bobby Allison scored with a Holman & Moody Mustang at a combined Grand American and Grand National event in 1971; a win that the sanctioning body refuses to recognize). Richard Petty suffered a potentially career- (and life-) ending wreck of his own in the Daytona 500. Although his #43 Pontiac was totally demolished, Petty was lucky enough to walk away to race another day.

Chevrolet drivers scored fewer wins than their archrival Ford foes for the first time in seven years during the 1988 season. Bill Elliott accounted for six of the nine Thunderbird wins recorded that year; those wins, combined with the nine other top-five berths he recorded, resulted in his first Winston Cup driving title. That title was the first for a Ford driver since the corporation had withdrawn from racing after the 1969 Grand National season, and it was the beginning of the heightened competition between Fomoco and GM drivers that has characterized the Winston Cup series ever since.

Pontiac driver Rusty Wallace scored six wins for the Indian-head division in 1988 to come in second behind Elliott in the points race. All told, Chevy drivers won eight times, with Pontiac drivers tallying the same, and Buick and Olds drivers each scored two.

NASCAR abandoned all semblance of "stockness" in 1989 when it allowed GM to campaign the Lumina body style. Though never UAW equipped with either a V-8 or rear wheel drive, that's just how the new Luminas raced—and won!—that year. *John Craft*

1989: Luminas Take the Torch

Rusty Wallace used his second-place finish in the 1988 points race as a springboard to the 1989 title. But before he got to walk across the stage in New York as the new Winston Cup champ, Darrell Waltrip crowned a stellar career with a win in the 1989 Daytona 500. His #17 Tide Monte Carlo SS crossed the stripe first in his 17th trip to Daytona.

En route to the championship, Wallace scored all 6 of the wins in the Pontiac column in 1989, while Chevrolet pilots Dale Earnhardt, Waltrip, Geoff Bodine, and Ken Schrader conspired to score 18 wins for Chevrolet. The most notable of those (Daytona 500 aside) was the Southern 500 cinched by Dale Earnhardt in the Goodwrench Monte Carlo. Ricky Rudd recorded a win for Buick at Sears Point, while Olds driver Harry Gant performed similar honors for that marque at the spring Darlington race.

Chevrolet drivers reluctantly had to surrender the "keys" to their tried-and-true Monte Carlos late in 1989 due to GM's decision to retire the marque in favor of the new and slipperier-shaped Lumina line. There was just one problem with that plan: Chevrolet wasn't building any V-8-powered, rear-wheel-drive Luminas. And that drivetrain configuration had been the rules-required standard since the beginning of NASCAR in 1949. But, of course, the NASCAR rules book has always been subject to change—at a minute's notice——when the sanctioning body is so inclined. And that's just what happened in 1989.

In many ways, the last links with the old Grand National days and the Strictly stock series that preceded it were severed when the Chevrolet Lumina was legalized for Cup competition. Certainly all pretense that NASCAR racing was in any way associated with "stock" production American sedans was forever ended in 1990.

That having been said, Chevrolet drivers wasted little time in proving that their new Luminas were every bit the race cars that Monte Carlo SSes had been.

So the on-track battles continue into a new decade.

Ernie Irvan
Kodak
Gold film
4
Delco Remy
AC
Winston
UNOCAL
Tyson
BUSCH
MOOG
GOODYEAR
STP
BILSTEIN
CRANE Cams
MOROSO
SPEED-PRO
Prestone
SIMPSON
RIGHT GUARD
Bowman
RICHMOND GEAR
Shurtape

Chapter 5

THE 1990s

A GLIMPSE OF THE FUTURE

The last decade of the century was destined to be a winning one for GM adherents. New and evermore aerodynamic body styles, new ponies squeezed old small block motor, and a string of Chevrolet biased rule book concessions all added up to NASCAR wins—a lot of them.

1990–1991: Luminas Power into the New Decade

The first full season for the all-new Lumina was in 1990, and Derrike Cope was the first to score a Lumina win in the Daytona 500. Dale Earnhardt was snake bit as usual in the 500 that year. He seemed destined for certain victory until he blew a right front tire on the last lap of the event. Ironhead went on to overcome that disappointment during the rest of the season, and by year's end he secured his fourth Winston Cup crown. His 9 wins (including a third Southern 500 triumph) helped Lumina drivers visit victory lane a grand total of 13 times. Pontiac drivers turned in 3 wins, while Olds and Buick drivers scored 1 each. Brett Bodine's win at North Wilkesboro in the spring race proved to be the last earned by a Buick driver in the Winston Cup series to date. Perhaps the marque will one day return to the starting grid on the NASCAR tour, but until it does, Buick's total of 56 series wins and 2 Winston Cup championships will remain unchanged.

California native Ernie Irvan burst onto the scene with his first Cup win in 1990 and backed it up by joining the very exclusive club of Daytona 500 winners in 1991. His Lumina win was the first of 11 scored by Chevy teams that season. Earnhardt accounted for 4 of those, and though others accounted for more, Ironhead's consistency (and 21 total top-10 finishes) earned him a fifth national driving title. He again missed out on a win in the Daytona 500 that year but made up for it with a superspeedway win at Talladega. Harry Gant put Olds out in front in the Southern 500 at Darlington. Four other Cutlass victories made 1991 the best year for Olds drivers on

When Chevrolet elected to retire the Monte Carlo marque (temporarily) in late 1989, Bow Tie teams had to switch over to Lumina-based comp cars. Though based on staid little front-drive, six-cylinder street cars, the rear-wheel drive, small-block-powered Luminas made fine race mounts for drivers like Ernie Irvan, Dale Earnhardt, and Darrell Waltrip.

the tour since 1979. Pontiac drivers recorded three wins of their own.

1992: Oldsmobile Signs Off

In retrospect, Davey Allison's win in the 1992 Daytona 500 (with a Robert Yates-prepped Ford) was probably an omen. And that's because 1992 was to become only the second year in the modern era (to that date) when a non-GM- based team was able to win the prestigious Winston Cup driving championship. Alan Kulwicki was the man to beat in the points race that season and his "Underbird" independent team overcame incredible odds to cinch the title for Ford.

Dale Earnhardt provided some reason to cheer with a win in the World 600 at Charlotte, and Ernie Irvan drove his Kodak Lumina across the stripe first at Daytona in the Firecracker 400 and at Talladega in the Talladega 500. In addition, Darrell Waltrip scored a popular Chevrolet win in the Southern 500 that season. Pontiac drivers also won at Richmond and Rockingham. Unfortunately, King Richard Petty scored none of those wins during his final season on the tour. It's a fair prediction to say that his record of 200 wins will never be bettered. It's just too bad that he couldn't have added an additional victory or two during his 1992 "Fan Appreciation Farewell Tour."

The two Oldsmobile wins scored by Harry Gant at Dover and Michigan turned out to be the last recorded by Cutlass drivers in the century. Like Buick, Olds elected to drop its racing funding following the 1992 season,

Richard Petty's last full season on the Winston Cup tour was spent behind the wheel of a svelte little Grand Prix like this one.

Though Richard Petty's driving days are in the past, his familiar #43 Pontiac is still a regular fixture on the Winston Cup tour. His most recent driver is John Andretti.

so GM teams for 1993 had to select between either Chevy or Pontiac sheet metal (for in reality, modern NASCAR "stock" cars are identical beneath the skin).

1993: Pontiacs Push Ahead

Dale Jarrett got 1993 off to an auspicious start for Bow Tie drivers by putting Joe Gibbs' Lumina in the winner's circle at the Daytona 500. Jarrett, one of a growing number of second-generation NASCAR drivers, did what his dad (Ned Jarrett) never got a chance to do during his own standout Grand National career—win the Daytona 500. Earnhardt reprised his now familiar role of bridesmaid in the 500 by finishing second to Jarrett, a mere 0.16 seconds behind him on the track.

Rusty Wallace gave an early indication of the strength that Pontiac drivers would display during the season by winning the second race on the 1993 tour at Rockingham in his Penske-prepped Grand Prix. All told, Pontiac drivers visited victory lane 11 times that year. And that was more than both Chevrolet and Ford drivers could say at the end of the 30-race season. Chevy drivers did make headlines with wins on big tracks like Talladega in the Winston 500 (Ernie Irvan), Darlington in the Rebel 500 (Earnhardt), Charlotte in the World 600 (Earnhardt), Daytona in the Firecracker 400 (Earnhardt), and back at Talladega in the fall race there (Earnhardt).

Although the Southern 500 was claimed by Mark Martin in a Ford and Ernie Irvan defected

Skittles
PONTIAC
MOOG
SELECT
SIMPSON
Nomex
36
76

Ernie Irvan drove one of the flashiest Grand Prixes on the circuit. His candy sponsorship is an example of the increasing number of non-automotive sponsors that have made an investment in Winston Cup racing.

mid-season to replace Davey Allison (tragically killed in a helicopter accident) in Robert Yates' T-Bird, on balance both Pontiac and Chevrolet drivers had a good year in 1993, which included winning the series title. Earnhardt claimed his sixth seasonal title with 6 wins and 17 top-five finishes, bringing him one notch closer to King Richard Petty's record 7 Winston Cup titles. Ironhead got to pocket $3,353,789 in compensation for those labors in addition to adding yet another trophy to his Lake Norman, North Carolina mantel.

1994–1999: Gordon Takes Center Stage

It might have been easy in 1994 to think that things would follow pretty much the same pattern for the rest of the 1990s: Earnhardt and Wallace dividing up wins, leaving the leftovers to hapless Ford pilots like Mark Martin. But that's not how things were destined to turn out. And the first evidence of that fact made itself known in 1994. For that was the year that a young lad from Indiana first got a chance to visit a stock car victory lane. Jeff Gordon was the youth in question, and his first Winston Cup win came on May 29, 1994, in the always-grueling World 600 at Charlotte. He soon became a regular fixture in victory lane.

Sterling Marlin scored yet another Chevrolet win in the Daytona 500 in his Kodak yellow #4 Lumina. Unfortunately for Chevy fans, Ford drivers won the next three races on the tour. It wasn't until Darlington that a Bow Tie driver (Earnhardt) was able to best the newly resurgent Fomoco forces. The new-found speed that GM defector Rusty Wallace and the rest of his T-Bird stablemates displayed in 1994 made

Dale Earnhardt won his fourth, fifth, sixth, and seventh Winston Cup championships with the assistance of Chevrolet Lumina race cars.

Earnhardt's chances of winning his record-tying seventh Winston Cup title more than a little problematic. This was especially so since by season's end Thunderbird drivers had scored nearly twice as many wins (20) as their Chevrolet rivals (11). Pontiac drivers fared even worse with hotshoe Rusty Wallace off at Ford and were shut out of the winner's circle completely. But despite all the obstacles, Earnhardt's consistency and the NASCAR rules book's preference for total top fives over total wins produced yet another Winston Cup crown for Earnhardt and team owner Richard Childress.

Change was the theme for 1995. First and foremost for Chevy drivers was the reintroduction of the Monte Carlo nameplate to the starting grid. Like the Luminas it replaced, the new-for-1995 Monte Carlo was based on a street-going version that sported front-wheel drive and a V-6 motor. The sanctioning body deemed the chassis "legal" for competition and even permitted Chevrolet aerodynamicists more than a little dramatic license with the new race car's hindquarters in order to create "parity" with Ford drivers out on the track. And that turned out to be a good thing for Chevy drivers.

Parity as defined by the sanctioning body that season translated into a Chevrolet sweep in the Daytona 500 (led by Sterling Marlin) and the next 6 races on the tour. It wasn't until April that a non-Bow Tie-badged race car got to grace victory lane. And even then, that win turned out to be the exception for 1995 rather than the rule. All told, Chevrolet

Though Jeff Gordon's boy-next-door looks might get him carded at a nightclub, he's been "The Man" on the NASCAR tour almost since his first Winston Cup race. His rainbow-hued Monte Carlo is easy to pick out as it effortlessly slices through the pack during the last laps of a race.

GOODYEAR
DieHard
CRANE Cams
76
MOROSO
TOOLS
VICTOR
SPEED-PRO

Derrike Cope
10
True Value
Winston
BUSCH
MOOG
Goody's
SIMPSON
Auto Meter
Gatorade
Stant
RIGHT GUARD
PERFECT CIRCLE
Bowman
altima
MSD
IGNITION
MAC TOOLS
QUARTER MASTER
PENNZOIL

Slight of build and high pitched in voice, Jeff Gordon is the very antithesis of the burly man's-man drivers who piloted GM race cars during NASCAR's earliest days. Be that as it may, the lad has left all of those two-fisted types in the dust on his tear-through the record books. *John Craft*

Previous pages
Derrike Cope in his Lumina was the surprise winner of the 1990 Daytona 500 when Dale Earnhardt's Lumina blew a tire coming out of turn four on the last lap of the race.

Monte Carlo drivers recorded 21 wins that season (to Ford's 8). Parity indeed! Jeff Gordon added 7 wins to the 2 he'd scored in 1994. Those triumphs earned him the nickname "Boy Wonder" and his very first Winston Cup driving title.

Slight of build and diminutive in stature, the youthful Gordon seemed to be an unlikely candidate for NASCAR stardom—especially when compared to "man's man" types like Curtis Turner, Fireball Roberts, Buddy Baker, and Dale Earnhardt. But that's just what he became—seemingly overnight—a NASCAR star. And so much so that the Winston Cup tour has become the Hoosier's playground during the second half of the 1990s.

As the new millenium dawns, young Gordon has won a grand total of three national driving titles (1995, 1997, and 1998) and in just six short seasons has recorded an incredible 47 victories (as of July 1999)—a figure that ranks him 11th on the all-time wins list.

In 1998 alone Gordon scored so many wins (13) that race promoters began to worry about a fan backlash similar to the one that occurred during the 1955 and 1956 series when Karl Kiekhaeffer's all-conquering Chryslers won just about everything in sight. So dominant has Gordon become that Ford and Chevrolet partisans alike howl in glee on those rare occasions when his pretty rainbow-hued race cars fall out of an event short of the winner's circle. Fellow Monte Carlo driver and Hendricks teammate Terry Labonte turned out to be the only match for Gordon during the last half of the 1990s, and Texas Terry proved that point by winning his second Winston Cup crown in 1996.

In the two seasons since, it's been all Gordon and just about all Chevrolet. Even the introduction of new jellybean-shaped (four door!) Ford Taurus race cars in 1998 wasn't enough to slow the Chevrolet juggernaut. As this book is being written, Chevrolet engineers have an all-new and even sleeker Monte Carlo racer just waiting in the wings to start racking up NASCAR wins.

Regardless of what Ford's sales literature might say, it's Chevrolet racers who are truly racing into the future! GM drivers have been around the NASCAR scene since the early days of NASCAR back in 1949. And they've been winning ever since. Don't look for that to change any time soon.

"Texas" Terry Labonte won his second Winston Cup championship with his Monte Carlo in 1996. But despite Labonte's win that year, most of the last half of the 1990s belonged to Jeff Gordon, who drove his Monte Carlos to championships in 1995, 1997, and 1998.

INDEX

Car Models